THERMAL ENERGY FROM THE SEA

THERMAL ENERGY
FROM THE SEA

Arthur W. Hagen

NOYES DATA CORPORATION

Park Ridge, New Jersey   London, England
1975

FOREWORD

This is another Noyes Data Technology Review which extends the study of oceanography into the search for alternate energy sources. Most of the information presented in this book is based on studies conducted by industrial and engineering firms or university research teams under the auspices of various governmental agencies.

Here are condensed vital data that are scattered and difficult to pull together. Experimental equipment and structures are reviewed and detailed by actual case histories.

Advanced composition and production methods developed by Noyes Data are employed to bring our new durably bound books to the reader in a minimum of time. Special techniques are used to close the gap between "manuscript" and "completed book." Industrial technology is progressing so rapidly that time-honored conventional typesetting, binding and shipping methods are no longer suitable. Delays in the conventional book publishing cycle have been bypassed to provide the user with an effective and convenient means of reviewing up-to-date information in depth.

The Table of Contents is organized in such a way as to serve as a subject index and provides easy access to the information contained in this book. A valuable list of bibliographic references is also given.

CONTENTS AND SUBJECT INDEX

AN OVERVIEW OF SEA THERMAL POWER   1
  Site Analysis   8
  Cost   14
  Systems Analysis   17
    Assessment of Ocean Thermal Gradient Systems   17
    Problems Areas   21

OPEN CYCLE OCEAN THERMAL GRADIENT POWER PLANT   27
  Preliminary Design Investigation   31
    Overall System Design   31
    Design of the Spray Evaporator   31
    Design of the Turbine   35
    Deaeration Losses   36
    Condenser Design   37
    Sizing of the Cold Water Pump   38
    Fresh Water Production   38
    Summary and Conclusions   38
  Feasibility Study of a 100 Megawatt Plant   39
    Design of the Turbine   42
    Falling Film Evaporator Design   43
    Deaeration of the Ocean Water   46
    Condenser Design   48
    Sizing of the Cold Water Pump   49
    Fresh Water Production   50
    Summary and Conclusions   50

CARNEGIE-MELLON UNIVERSITY DESIGN   52
  Overview of the Design   55
    Water Circulation System   55
    Working Fluid Circulating System   55
  Seawater Inlet-Outlet Hydrodynamics   58
    Topology of SSPP   58
    Relation of Plant Size and Seawater Flow 58
    Two-Layer System: Model for Warm Water Intake   58
    Intake Water Structure   60
  Design Features   61
    Boiler Technology Considerations   61

Cold Water Pipe vs Ammonia Pipe   67
Cold Water Intake Pipe   69
Water Pump   72
Gas Turbine   77
Antifouling Cost   77
Vertical Tube Heat Exchangers   79

UNIVERSITY OF MASSACHUSETTS DESIGN   81
Overall Design Concept (Mark I)   81
    Cold Water Supply   84
    Cold Water Suction Pipe   85
    Turbine and Turbine System   86
    Naval Architecture   86
    Anchor and Mooring System   87
    Energy Umbilical   88
    Control of Biofouling   89
Heat Exchanger Design (Mark I)   89
    Exchanger Requirements   90
    Materials   96
Variations in Heat Exchanger Design (Mark II)   99
    Condenser Placement   100
    Boiler Ocean Side Pumping System   102
    Heat Exchanger or Power Module Size   103
    Site Location   105
    Heat Exchanger Fouling   106
    Passage Size   106
    Cladded Heat Exchangers   106
Turbine Design (Mark I)   107
Technical and Economic Evaluation of Turbomachinery (Mark I)   111
    Turbine Size   111
    Turbine Cost   117
    Generator Cost   121
    System Losses   121
Technical and Economic Feasibility (Mark II)   121

COMPARATIVE STUDY OF CLOSED CYCLE TECHNOLOGIES   127
Cost Comparison of CMU and UMASS Designs   127
Heat Exchangers   128
Working Fluids   132
Turbines   132
Cold Water Supply and Pipe   133
Mooring and Anchoring   133

ADDITIONAL TECHNOLOGIES   134
Hydrogen Utilization   134
Hydrogen Energy   136
Nitinol Utilization   141
Further SSPP Proposals   144
    Ocean Thermal Energy Conversion (OTEC)   144
    Danish System   145
    Japanese System   146

REFERENCES   149

AN OVERVIEW OF SEA THERMAL POWER

The information in this chapter is excerpted from the following publications:

NASA TM X 70783
AD 179 877
PB 228 066
PB 231 142

The idea of extracting useful work from a heat engine powered by the temperature difference between warm surface water and cold, deep seawater has received
some attention since the late 19th century, but a large scale power plant has yet
to be built and operated successfully. The basic principles of operation for the
latest concept of such a plant are illustrated in Figure 1.1. The figure shows a
proposed sea thermal power plant operating between seawater temperature levels
of 5° and 25°C. The entire plant is neutrally buoyant and is submerged at a
depth of 60 meters.

Since a seawater temperature difference of at least 15°C should be available to
operate such a system, there are a limited number of sites where thermal plants
can operate without excessively long pipelines. Both coasts of Africa, the west
and southeastern coasts of the Americas, and many Caribbean islands are situated
where the seawater decreases from a surface temperature of 25° to 30°C to 4° to
7°C at a depth of about 750 meters. Thermal power plants and associated systems for desalination and mariculture have been proposed recently by several
groups to take advantage of these suitable locations.

In 1881 Jaques d'Arsonval published his thoughts on the possibility of constructing a steam power generator that made use of the temperature difference between
the surface water and the deepwater in the tropical seas. In 1930 Georges Claude
actually constructed this type of Solar Sea Power Plant (SSPP) at Matanzas Bay,
Cuba. This plant had a 22 kwₑ output. This experimental SSPP operated by
taking advantage of a 14°C thermal gradient. The cold water was brought up from
700 meters below the sea surface in a 1.6 meter diameter pipe which extended
to 2 km offshore. The plant operated for two years (1929 and 1930). In
the 1950s a French corporation, Energie Electrique de la Cote d'Ivoire, planned

1

FIGURE 1.1: SEA THERMAL POWER PLANT

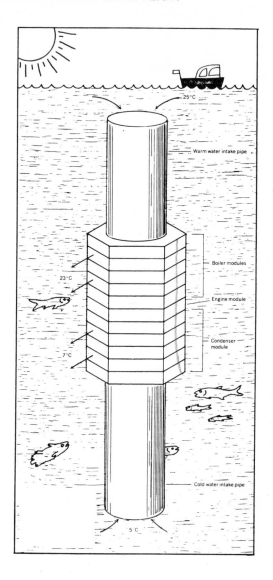

Source: AD 779 877

the construction of an SSPP at Abidjan Ivory Coast. This plant was never put
into operation. This SSPP was to have produced steam power based on a 20°C
thermal gradient. The cold water would have been brought up from a depth of
430 meters 4 kilometers offshore and, pumped through a 2 meter diameter pipe.
The flow through the cold water pipe was designed to operate at 2 m/sec. However, approximately 25% of the power produced by this SSPP would have been

consumed by the operations of the SSPP itself. This plant never reached full operation although several of the subsystems were built and installed. A floating plant to extract power from the oceanic temperature differences in equatorial regions, using propane as the working fluid in a closed system, has been proposed.

In the last five years the primary interest has centered on harnessing the temperature gradients in the sea to produce power. Despite the abortive attempts by Claude in Cuba and by the French group in Nigeria to generate power from the temperature difference between warm and cold seawater, recent advances in heat transfer surface design and thermal power plant design have suggested that sea thermal power is now competitive with more conventional generating methods. Theoretically, work can be done by a heat engine so long as a temperature difference exists between two heat reservoirs (in this case warm and cold seawater). The maximum thermodynamic efficiency of such a system is

$$\eta = \frac{T_2 - T_1}{T_2}$$

where T_2 is the temperature of the hot reservoir and T_1 is the temperature of the cold reservoir in absolute degrees. This temperature difference is hundreds of degrees in efficient heat engines and turbines, but only about twenty degrees in seawater. In a typical example the warm water is at 30°C and it is cooled to 25°C to generate low pressure steam at the latter temperature. The steam is condensed at 15°C by heating the cold water from 5° to 10°C so that the high and low temperatures for the steam cycle are 25°C or 288°K (absolute). The maximum thermodynamic efficiency is then

$$\eta = \frac{298°K - 288°K}{298°K} = .033 \text{ or } 3.3\%$$

When account is taken of the losses in an actual system, the efficiency obtained is likely to be only about 2% or even less. This low thermal efficiency does not appear to be a fatal drawback for such a device. A conventional coal generating plant has an overall efficiency of about 30% as compared to the 1 to 3% value given above, so that a sea thermal plant must transfer about ten times as much heat to yield the same power output.

However, the sea thermal boiler is designed to operate at a low pressure differential and ambient pressure and does not need the complex vessel required to operate at both high pressure and temperatures. Thus the less rigorous operating regime of the sea thermal plant is expected to bring the boiler-heat exchanger costs into a range which is competitive with high pressure steam boilers.

An early design by Claude, off the coast of Cuba, utilized a low pressure flash evaporation chamber to produce steam, followed by a turbine, and finally a direct contact condenser using cold, deep seawater. Claude's attempt failed for a number of reasons, but chiefly because of the use of water itself as the working fluid. This avoided heat-transfer problems but created others of great difficulty including that of the corrosiveness of seawater. At low temperatures, the specific volume of steam is extremely large. This means a large turbine and, with low Reynolds number, low efficiency. The corresponding high vacuum required the bothersome necessity and expense of removing dissolved gases from the water and the maintenance of large leaktight connections. Another chief reason for failure was because the land-based plant necessitated extremely long lines to bring water from the depths. With long lines (2 km long) and a small plant (40 kilowatts), heat flow into the cold lines turned out to be excessive. The 7000 kilowatt plant in Nigeria was based on this design also and full operation was never realized.

More recent efforts have eliminated the major faults of Claude's design. Two primary examples are: (1) controlled flash evaporation (CFE) of steam for the production of power and fresh water; and (2) a closed cycle plant using a suitable working fluid to drive a power turbine [cf. Figure 4.1 (UMASS Design)].

In the CFE process, seawater at the warm ocean surface temperature of about 25°C enters the evaporation chamber as shown in the upper left of Figure 1.2. The process is described as follows. In controlled flash evaporation, seawater at the top passes through orifices between orifice spacer blocks. As the pressure lowers, water evaporates and cavitation forms a core stream of vapor, with a film of liquid descending on each side wall of the chute.

Vapor separates free of seawater in the lower expanded section. The evaporating conditions are shown at left at respective distances of passage of the liquid. Steam generated passes through back to the expansion turbine to generate power, thence to a surface condenser for condensation by cold deep seawater.

The process is further described wherein a combined system for mariculture, fresh water, and power production. An actual flash evaporator in a full scale system would contain many parallel evaporation chambers similar to the single unit shown in Figure 1.2. The CFE process has overcome the problem of removing dissolved gases from the water and has minimized corrosion problems. It also minimizes temperature losses between the seawater and vapor, and brine entrainment.

Power can also be generated by using in a closed cycle a suitable fluid with good heat transfer characteristics and a high vapor pressure at ambient temperature. Among the candidates suggested are ammonia and propane. A typical closed cycle power plant designed to use the temperature difference between warm and cold seawater for its operation [Figure 4.1 (UMASS Design)] utilizes a closed Rankine power cycle made up of a turbine, pump, condenser and boiler. The working fluid (propane is favored in the UMASS concept) is evaporated in the boiler and power is generated by expansion in the turbine. After leaving the turbine the low pressure vapor is condensed in a heat exchanger using cold seawater and is pumped back to the boiler pressure.

A similar system using ammonia as the working fluid is described in the Carnegie-Mellon University (CMU) concept. Both studies propose to submerge the plant in the water to a depth of 30 to 60 m in order to employ the water's hydrostatic pressure head to compensate for the vapor pressure of the working fluid in a boiler.

Sea thermal power plants such as the two just discussed are, of necessity, very large in size because of the seawater throughput rates. CMU estimates that the mass flow rate of the boiler would be roughly equal to the quantity of water passing through a typical hydroelectric plant. The plant proposed in the UMASS study for a location 25 km east of Miami in 360 meters of water is approximately

150 meters wide, 205 meters long and 120 meters high. The cold water inlet tube, reaching almost to the sea floor, is about 325 meters long. A line drawing of the proposed system is shown in Figure 4.1 (UMASS Design). One interesting aspect of the University of Massachusetts plant is that it is designed to be moored in a nominal 2 knot current, thereby transferring heat in the propane boiler by a forced convection throughput of warm seawater. Fouling of the heat transfer surfaces as a result of immersion in tropical seawater is mentioned as a potential problem for the design.

FIGURE 1.2: CONTROLLED FLASH EVAPORATION (CFE) PROCESS

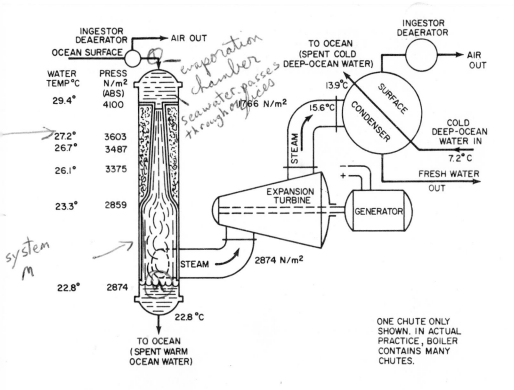

Source: AD 779 877

The CMU study proposed to use the power generated by a thermal difference plant to generate hydrogen fuel by electrolysis. This would do away with a costly network to distribute the generated power to land-based consumers as electricity. A scenario in which offshore generated electric power is used to electrolyze seawater and the resulting hydrogen and oxygen gases are then distributed in pipeline networks has been dubbed the hydrogen economy. A diagram of this future system is given in Figure 1.3.

FIGURE 1.3: PROPOSED HYDROGEN ECONOMY

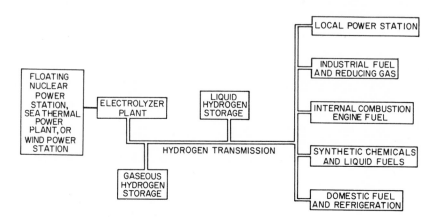

Source: AD 779 877

Environmental Impact: Since a sea thermal plant operates by removing heat from surface water and by heating large quantities of cold, deep ocean water, questions have been raised as to the possible deleterious effects of constructing large numbers of sea thermal plants in equatorial waters, considering the fact that heat is transferred from the upper ocean to the deep layers with a slight decrease in surface temperature. This lower surface water temperature results directly in a corresponding decrease in the atmospheric temperature near the surface.

If the incoming solar radiation remains constant, then the reduced surface temperature results in less evaporation and outgoing radiation and a greater net heat transfer to the water. The result is presumably increased convection away from the tropical plant location. More detailed studies of potential environmental effects are necessary for sea thermal power plants, but it appears that the overall environmental impact of these plants would be within acceptable bounds.

Cost: The cost of a closed-cycle sea thermal plant of 100 Mw capacity was estimated in detail in 1972. Based on an estimated 1972-1973 cost, the sea thermal plant is competitive with either nuclear or coal power. An additional cost estimate for a sea thermal power plant is listed in Table 1.1. Heat transfer surfaces have recently been developed which would significantly decrease the capital cost of a sea thermal plant and place such a facility in an even more favorable position.

The proposal has been made to use sea thermal power to electrolyze seawater and to produce hydrogen, and both estimates in Table 1.1 include a figure to take account of the cost involved in the electrolysis process and the handling and distribution of the hydrogen fuel. An alternate proposal has been made for a 7180 kw net power output open-cycle steam plant which yields as a by-product 6 million U.S. gallons of fresh water per day at a total installed cost of

$18.4 million. For one set of optimum design conditions, electric power was projected at 6 mills/kwh and fresh water at $1/1,000 U.S. gallons.

TABLE 1.1: AN ESTIMATED COST OF SEA THERMAL POWER GENERATION*

	- - - - Megawatts- - - -	
	100	400
Capital cost ($/kw)	390**	570***
Fixed charges (mills/kwh)	8.4	12.2
Fuel cost	–	–
Operation and maintenance costs	0.9	0.9
Total power cost (mills/kwh)	9.3	13.1

*Load factor is 0.8
**Capital cost includes $220/kw plant cost, $170/kw hydrogen
fuel generating, storage and distribution system to shore.
***Capital cost includes estimate of $400/kw plant cost, $170/kw
hydrogen cost

Note: Fixed charges (FC) = $\dfrac{\text{Amortization cost (AC)} \times \text{Capital cost (CC)}}{\text{Load factor (LF)} \times 8760}$

Amortization cost = 15% for above estimates.

Source: AD 779 877

Mariculture: The SSPP concept offers, in addition to its power production capabilities, the gift of mariculture. One may define mariculture as ocean farming. Mussel farming has proved very successful in Spain and the Japanese have large oyster culture operations. The farming of shrimp is becoming commonplace and fish have been raised for centuries. The selection of breeding stock for mariculture is a well established science.

For an adequate period of time an algae growth plant (farm) was proved successful. Algae (*Chlorella pyrenoidosa*) growth rates of 14.4 g of dry algae per square meter per day were achieved. In this experiment a nutrient medium was used. The nutrient medium associated with the SSPP is the ocean itself. Deep ocean water (the cold water used for condensing purposes) is many times richer than surface water in nitrates, phosphates and silicates, which are essential nutrients for phytoplankton growth.

The SSPP condenser (nutrient rich) water can be routed into lagoons, tanks or ponds where it will stimulate the photosynthetic process by providing the necessary nutrients. The phytoplankton concentration in the mariculture system may then be used to obtain optimum yields of shellfish, shrimp larvae, filter feeding fish, etc.

Efficiencies of 70% have been obtained for Euphausia Pacifica feeding on Dunaliella. Efficiencies of 55% for brine shrimp, Artemia, feeding on Dunaliella have been obtained. However, assume only a 10% rate of phytoplankton conversion to a secondary producer. This 10% conversion operating on a single acre would yield 20 tons of secondary producer per year. Note that mariculture can also be accomplished in the open sea by the artificial SSPP upwelling system. A site 1.6 km offshore from St. Croix where a 1,000 meter ocean depth was found to occur was chosen.

It was found that the deepwater, in addition to its nutritious values, was free of human disease producing organisms, shellfish parasites, predators, fouling organisms and man-made pollutants. This water was pumped onshore and grew clams, oysters and scallops to market size in a year.

SITE ANALYSIS

Although the technical feasibility for production of electricity from the energy potential associated with temperature gradients in ocean water has been elucidated, the practical accomplishment of economically beneficial power production has not been demonstrated. Because construction of a power production facility will utilize some new materials and fabrication techniques and must consider design problems associated with incompletely assessed hazards of the marine environment and material incompatability associated with use of some kinds of working fluids, the economy of production of electrical energy or elemental fuel, i.e., hydrogen, from seawater temperature differences is tenuously assessable.

A prototype facility should be designed and put into operation at a locality with the greatest potential for benefit from the facility with a maximum of economy. Because the production of power from seawater temperature differences could produce marketable commodities such as fresh water and the attendant utilization products of this water in aqua culture, the economic potential of a prototype power production facility should be assessed in terms of utilizing all attendant products of the plant facility.

In addition, because the maximum of energy potential of a power plant system is directly related to the magnitude of temperature differences in seawater and the distances and volume of water to be moved through the facility, a site for the prototype facility was sought where large temperature gradients are available. The market for fuel or electrical power and by-products from power production must be near the plant facility for maximum economic benefit to be derived from the plant operations.

A prototype power production facility designed and in operation at any location should provide useful technical information for subsequent facilities' construction at other locations, recognizing that different possible locations for a plant facility could involve wide differences in facility design. For example, a power production facility located in the Gulf Stream would involve the use of a substantial platform compared to a platform needed for a nearshore plant. An onshore plant would not need a platform on which to put the plant but could involve water piping over greater distances than needed for the deepwater Gulf Stream facility.

Environmental considerations in design of a solar sea power plant are readily identifiable in general; however, some plant locations could have unique characteristics which would be identified once a site is specified. Environmental parameters to be analyzed at each tentative site for a prototype facility are listed below. Because the atmosphere and ocean circulations undergo long period, broad scale changes it is probable that most locations selected for a solar sea power plant will be affected by the results of the circulation changes. Whether or not the changes are of any consequence to plant operations or plant economics depend on final plant design and location.

The near proximity of warm and cold water bodies (not considering geothermal hot water sources) is frequently produced by ocean current or upwelling circulations. These circulations are produced by surface wind stress, and with respect to upwelling, coriolis acceleration of wind driven surface water. Through atmospheric teleconnection with ocean circulation, variation in atmospheric circulations is reflected in variation of ocean circulation and vice versa. Periodic long-period oscillations in atmospheric circulation over the tropical Pacific oceans have been found to be directly related to water temperature variations in the tropical ocean and it will be shown here that water temperature along the coast of the United States varies in a periodic way but partly out of phase with variation in low latitude atmospheric circulation.

Because the origin of the Gulf Stream is associated with water advected into the Gulf of Mexico, a change in the northeasterly trade winds in the tropical Atlantic will affect the Gulf Stream. In Figure 1.4, a graph of the size of subtropical anticyclones in the tropical Atlantic ocean area is shown for the decade of the 1960s. The data depicted here are the measured geographical ocean areas with monthly average surface air pressure of 1,020 mb or greater normalized by the standard deviation of the area coverage and smoothed over 12 months.

This graph indicates the existence of a long-period oscillation in the atmospheric circulation. Because the anticyclone size controls the pressure difference between the high pressure region in latitude of about 30° to 35° and the equatorial region, variations in the anticyclone are expected to produce variations in the trade winds. The seawater temperature data graphed on Figure 1.4 along with the anticyclone data are the water temperature anomaly for each month for the 2.5° x 2.5° area off the coast of Newfoundland on the approximate position of the Gulf Stream Front (area centered 47.5°N, 37.5°W).

These data are smoothed over 12 months and it indicates that changes in water temperature in the Gulf Stream are related to changes in the atmospheric circulation. The variation in phase between the water temperature change and the change in anticyclone size, i.e., short phase difference for high amplitude change in anticyclone and long phase difference for low amplitude changes, suggests that the explanation for temperature change at the ocean observation site is due to decreased current speed or temperature in the Gulf Stream rather than wander off the stream axis across the observation point.

In Figure 1.5, the monthly averaged surface water temperature measured at coastal sites on the United States West Coast, the South American West Coast and at Bermuda are graphed as water temperature differences. The differences graphed are the monthly average temperature from one month of one year subtracted from the average temperature of the same month of the past year.

Therefore, during times of decreasing temperature the graphed differences are positive. From these graphs it is demonstrated that ocean upwelling circulation, in these different ocean areas depicted, varies in phase together producing temperature changes of about ± 2°F over about a three year period. Ocean surface salinity changes which are attributed primarily to rates of surface evaporation or rainfall have been found to be about 3 to 4 parts per thousand in seasonal variation with additional change of 1 to 2 parts per thousand with long period changes in cloud cover and rainfall associated with the prominent biennial oscillation in the atmospheric circulation.

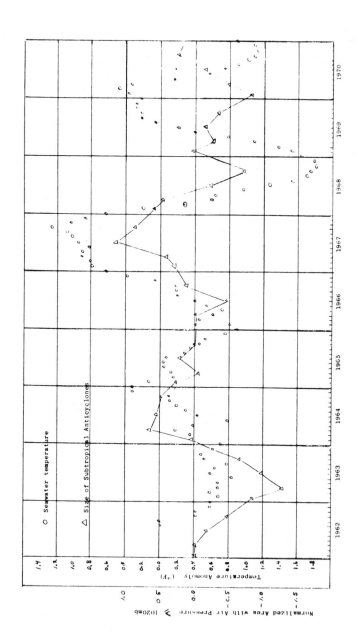

FIGURE 1.4: SIZE OF SUBTROPICAL ANTICYCLONES IN NORTH ATLANTIC OCEAN AND SEAWATER TEMPERATURE IN GULF STREAM NEAR NEWFOUNDLAND

Source: PB 228 066

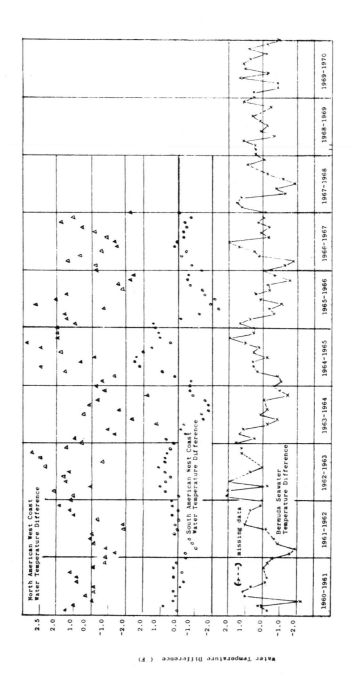

FIGURE 1.5: WATER TEMPERATURE VARIATION IN THE ATLANTIC AND PACIFIC OCEANS

Four factors should be used to judge the suitability of a site for a plant facility: (1) the suitability of multiple use of plant output, i.e., power, fresh water, aqua culture; (2) maximum gradients of water temperature and maximum differences in water temperature with high volumes of water supply; (3) proximity of market for plant products; and (4) social benefit.

The sites selected for a prototype facility are placed into two groups, one group being island sites and the second being deepwater ocean locations. In the first group the islands Puerto Rico, St. Croix and Hawaii were selected because all the selection factors are applicable to these sites and the successful operation of a prototype plant (1 to 10 Mw) would immediately provide significant benefit in terms of energy supply to the locale.

The second group of sites included the Gulf Stream and along the southern California coast. Because a significant portion of constructing a prototype plant for sites in this group would be construction of a platform on which to put a prototype plant the island sites seemed to be best suited for the prototype. In addition, the possibility that geothermal sources for warm water could be utilized in the Hawaiian region made that region most attractive for this program.

Listed below are both environmental and economic parameters that would be required for detailed analysis when a specific site for a prototype facility is delineated. Some of these parameters may be well-known at some of these sites while others will require special observation programs.

Economic Parameters for Site Analysis

Site availability
Market for products
Social benefit
Construction cost
Maintenance and operation cost
Environmental impact
Salvage value

Environmental Parameters for Site Analysis

Bottom survey
Water temperatures (time and spatial variations)
Currents
Wave, wind, tide
Geothermal (other sources of warm water)
Water quality

The political aspects of site selection may be considered in the economic parameter of site suitability analysis. Another question, political in nature, is the international legality of exploiting power in the free currents of the oceans and the potential hazard to shipping of a deepwater solar sea power plant.

There are two distinct SSPP types with respect to site selection: the shore-based system and the ocean-based system. The typical shore-based system would be located as described above on the coast lines of Florida, Puerto Rico, Hawaii, etc. However, the availability of suitable coastline constrains the gross quantity of power that shore-based systems can generate.

Ocean-based systems will typically be constructed so as to be a self-contained floating or tethered platform submerged below the surface of the sea, but still above the thermocline. The optimum location for such a plant is in tropical waters, 23°N to 23°S. Transmission and storage problems may limit the amount of power delivered by ocean-based plants.

For example, if SSPPs were to supply 10% of the U.S. energy requirements by the year 2020, about 0.3 million tons of hydrogen would have to be shipped each day. That is the equivalent of 30 shiploads per day where each ship carries 10,000 tons of hydrogen per trip. Underwater hydrogen pipelines might be feasible, or direct DC electric current transmission with DC to AC converters on shore, may be the most economical solution to the energy transportation problem.

The design of a land-based system is not as complex as that of an ocean-based system and both the CMU and UMASS SSPP concepts can be applied to land-based systems with little redesign of the working components. In fact, there has been a lot more experience with land-based piping systems that are applicable to SSPP design. The Ivory Coast pipeline was designed to be 4 miles long; there are many sewer outfall pipes longer than 4 miles made out of concrete. On Cyprus there is a 2 mile long plastic wastewater pipe that was installed by unreeling the pipe and simply anchoring it to the sea floor.

To obtain some insights into the siting problem, the Japanese experience can be considered. After studying the entire Japanese coastline it was concluded that while the water temperatures below 200 m are less than 10°C almost everywhere, the surface temperatures show great variation with the season. Situations of this type are also characteristic of much of the U.S. coastal waters. The relative attractiveness of the Pacific Islands such as Truk and Guam are, however, to be noted.

When considering the North Atlantic Ocean and the entire eastern seaboard of the U.S., significant temperature variations in the surface temperatures comparable to the Japanese situation are observed. The annual range of the Atlantic coast temperatures is greatest at the north end of the coast: 2.8° to 15.5°C, while the smallest temperature difference at the surface: 24.4° to 29.6°C is observed near Miami. This 5.2°C difference is important because the Miami site has been proposed as a potential SSPP site.

This temperature difference can greatly influence the efficiency of an SSPP and even cause an SSPP demonstration project to fail. The advantage of a tropical SSPP with a maximum temperature difference of only 2°C is notable. Truk and Guam are shown to be better SSPP sites than Miami on the basis of ΔT considerations. On the other hand, if one is constrained to build an SSPP near the U.S. mainland, Miami is the best site available on the east coast.

The surface temperatures during the critical month February indicate temperatures up to 24°C at Miami. The temperature differences at various places off the North Atlantic coast in February appear to be sufficient (marginally) to achieve a 1.25% efficiency. However, in August these temperature differences all but disappear. There is a gradual decrease in the temperature difference from February to August and a horizontal gradient SSPP would be efficient for only two months of the year. It is also interesting to note that the surface temperatures near the Gulf coast in February are about 15°C; insufficient to operate an

SSPP. Thus the only viable potential SSPP site off the coast of the U.S. appears to be the Miami site in the Gulf Stream or Florida current. That part of the Gulf Stream between the Strait of Florida and Cape Hatteras is usually called the Florida current. Another reason for selecting Miami as a prime site is the temperature behavior of the subsurface waters. Only off Miami does the temperature decrease with depth throughout the year, elsewhere it decreases only during the summer. At a depth of only 200 m one begins to find temperatures of 8°C at the Miami site. This site is recommended as the prime U.S. SSPP site.

It is recommended that the prime sites for SSPP construction be: Puerto Rico (south coast); St. Croix, U.S. Virgin Islands (north coast); Hawaiian Islands (various sites); the various U.S. Pacific Islands (Micronesia), largest sea thermal resource in the world, 5,000,000 square miles; and Miami, Florida.

In addition, there are numerous foreign sites; many in areas lacking electric power and fresh water. Any successful U.S. SSPP would tend to generate great demand by various nations for SSPP technology. The potential here is enormous not only because of the SSPP's power potential, but also because of the SSPP's food and fresh water production capabilities. The food producing capabilities of the SSPP will be discussed.

COST

The attractiveness of the SSPP (even without considering desalination and mariculture) is due to cost. The SSPP does not need acres of expensive solar collectors or expensive storage units for nights and cloudy days. The ocean continuously acts as a collector and as a heat storage unit permitting the system to be operated the year round, 24 hours per day. It has been estimated that a SSPP can be constructed for $200/kw (SSPP fuel is free).

Estimated costs are shown below. Estimated costs for electricity for an ocean thermal gradient power plant, based on fixed charge rate of 14.2% per year (without storage costs) are tabulated below.

Initial investment, power generation:
1980	$346 to $692/kw
1990	$390 to $780/kw
2000	$490 to $980/kw

Other costs:
1980 to 2000	$1,281 to $2,416/kw
1990 to 2010	$1,610 to $3,036/kw
2000 to 2020	$1,610 to $3,036/kw

Total 20 year cost:
1980 to 2000	$1,627 to $3,108/kw
1990 to 2010	$2,000 to $3,816/kw
2000 to 2020	$2,100 to $4,016/kw

Plant factor:
1980 to 2000	0.60
1990 to 2010	0.60
2000 to 2020	0.60

Average Price of Electricity at Plant:

1980 to 2000	15 to 30 m/kwh
1990 to 2010	19 to 36 m/kwh
2000 to 2020	20 to 38 m/kwh

Ocean thermal gradient power production has the potential for supplying 100% of the U.S. electric energy requirements in the not too distant future. By the year 2000 it could supply up to 5% of those requirements, and by the year 2020, 10% or more. To get some idea of the magnitude of the energy available consider that in 15 min the earth intercepts enough radiant energy from the sun to equal the amount of energy that is consumed worldwide in the form of fossil and nuclear fuels each day. In less than four days, an amount of energy equal to all of the earths fossil fuel reserves is intercepted. In order to evaluate the situation a bit further, the schematic plant design shown in Figure 1.6 was derived.

FIGURE 1.6: SOLAR SEA POWER PLANT SCHEMATIC

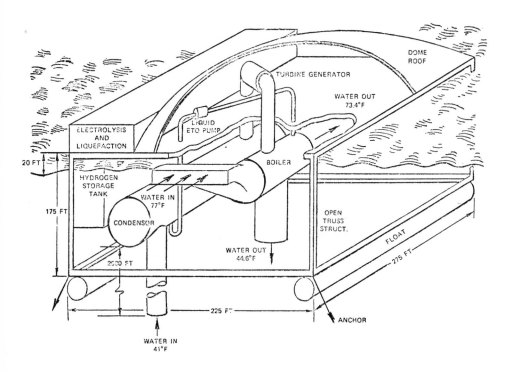

Source: PB 228 066

A rough estimate of the cost of the plant shown in Figure 1.6 was made for each of three proposed working media: ammonia, propane and ethylene oxide. The results (Table 1.2) indicate that a plant using ethylene oxide would cost a great deal more than one using ammonia. Use of propane would result in an intermediate cost.

TABLE 1.2: PRELIMINARY COST ESTIMATE FOR THE SSPP

Working Medium	Ammonia	Propane	Ethylene Oxide
Component	Cost (Millions of dollars)		
Boiler	4.2	4.3	4.3
Boiler feed tube	0.01	0.01	0.01
Boiler water collector tube	0.4	0.4	0.3
Condenser	7.5	13.8	10.6
Condenser exit water deflector	0.2	0.2	0.2
Condenser cold water pipe	3.4	3.2	2.9
Turbine piping	0.02	0.03	0.03
Working medium	0.24	0.20	0.39
Turbine generator	2.8	6.5	20.1
Liquid pumps	1.0	1.2	1.1
Platform	5.0	5.1	5.1
Miscellaneous piping and auxiliary items (10%)	2.5	3.5	4.5
TOTAL	$27.3 M	$38.4 M	$49.5 M
Net power output* (MW)	92.3	85.1	89.9
Plant cost per 100 MW (millions of dollars)	29.6	45.1	55.1

*Rankine cycle net power (100 MW) minus pumping power.

Source: PB 228 066

Table 1.3 shows the amortized plant costs for the three working media and the resulting electric power costs on a mill per kilowatt hour basis.

TABLE 1.3: COST OF ELECTRIC POWER AT SSPP PLANT

	Ammonia	Propane	Ethylene Oxide
Construction Cost, 100 MW Plants			
Initial cost ($M)	29.6	45.1	55.1
Amortized cost, 25 yr @ 8% ($M)	68.5	104.3	127.4
Amortized cost ($/kw)	685.0	1043.0	1274.0
Amortized cost (mill/kwh)	3.1	4.8	5.3
Insurance (1% of initial cost, mill/kwh)	0.3	0.5	0.6
Subtotal. fixed cost at 100% capacity (mill/kwh)	3.4	5.3	6.4
Operating costs. 30 men at* $33 K/m-hr (mill/kwh)	1.2	1.2	1.2
Total cost (mill/kwh)	4.6	6.5	7.6

*Arbitrary estimate – may be high.

Source: PB 228 066

These figures indicate a decided advantage to ammonia as a choice for the working fluid in a solar sea power plant. However, the selection of an optimum working fluid requires careful evaluation of many factors such as toxicity, materials compatibility, fire and explosive hazards, and plant maintenance and operational factors. Ammonia has problems in several of these areas, whereas propane would have fewer problems. Freons would be better still from a safety and handling standpoint, but plant cost would be higher than for propane.

SYSTEMS ANALYSIS

Assessment of Ocean Thermal Gradient Systems

Criterion	Assessment
(1) Possible Impact on Energy Requirements	
(a) Types of Products	Electricity; hydrogen fuel gas; oxygen
(b) Types of Users	Households, utilities, industries, ships (for re-fueling), etc.
(c) Siting Restrictions	Ocean-based systems might consist of a self-contained floating platform submerged to a depth of $\geqslant$ 200 feet below ocean surface, and moving in a defined orbit to collect warm surface water. Such systems would be located in deep ocean waters and mostly in tropical latitudes (i.e., 23°N to 23°S).

(continued)

Criterion	Assessment
	Shore-based systems, drawing warm water from the ocean surface and cold water from the ocean depths and would be located along the coast of Puerto Rico, Florida, California, Hawaii, etc.
(d) Potential Capabilities	By the year 2000 these types of systems might be able to supply over 1% of the total U.S. energy requirements (i.e., about 2×10^{15} Btu per year) and by the year 2020 about 10% of the U.S. energy requirements (i.e., about 18×10^{15} Btu per year).
	The availability of suitable coastline will constrain the quantity of power that could be generated by shore-based systems. Problems of transmission of electricity or storage and transportation of hydrogen might constrain the quantity of power that could be delivered by ocean-based systems.
(2) Technical Characteristics	
(a) Flexibility	Since the surface of the ocean acts as a medium for storing vast quantities of solar energy, ocean thermal gradient systems could be used to generate electricity and/or hydrogen, continuously, to respond to time-varying loads. No supplemental energy supplies would be required by the system.
(b) Technical Reliability	Reliability of this type of system might be affected by such factors as: marine fouling; corrosion of various parts of the system; strengths of materials, such as those used in the long insulated water-transport pipes required; as well as the reliability of the facilities for transmitting electricity or transporting hydrogen from an ocean-based system to shore points; or changes in the ocean currents used to remove mixed or used water from a fixed shore-base system.
(c) Sensitivity to Weather	Neither system would be affected by normal changes in weather. Ocean-based systems might be disrupted by severe storms or hurricanes, depending on system design and modes of operation. Exposed pipes of shore-based systems might also be subject to storm damage.
(d) Security	Exposed locations of ocean-based system sites might make these types highly vulnerable to sabotage or enemy action, especially outside of the 12 mile jurisdictional limit of the U.S.
(e) Base-Load/Peak-Load Capabilities	Shore-based systems could meet base-load and peak-load demands without storage. Ocean-base systems could do likewise, if suitable electrical transmission lines, to shore points, could be provided.

(continued)

Criterion	Assessment
(f) Storage Requirements	Ocean-based systems, that use electricity to generate hydrogen, would require at least temporary intermediate ocean-based storage capabilities. For instance, if this type of system were used to supply 10% of the U.S. energy requirements by the year 2020, about 0.3 million tons of hydrogen would need to be shipped or piped to shore points each day. (Shipping requires about 30 shiploads per day, using ships capable of carrying 10,000 tons of hydrogen per trip.) Total ocean-based storage capabilities might need to be at least several million tons of hydrogen, if ships are used for transport.
(g) Interfacing Requirements	Transmission distances are such that it would be preferrable to transmit DC electricity from ocean-based systems to shore points. If this is done, DC to AC inverters would be required at the shore facilities. Hydrogen, if shipped or piped to shore, would need to be converted to electricity by means of fuel cells and inverters, or by means of Aphodid burners (see U.S. Patent 3,459,953) and the AC generators.
(h) Redundancy	Shore-based and ocean-based facilities could be interconnected to increase total system reliability.
(i) Synergistic Effects	Ocean thermal gradient systems could be interconnected to intermittent types of solar energy systems, such as Photovoltaic, Wind Energy, Solar-Thermal, Process-Heat or Systems for Heating and Cooling of Buildings, in order to increase their performance and reduce their total storage requirements.
(j) Transportation, Transmission and Distribution Requirements	Shore-based systems would offer no more electrical transmission problems than present types of electrical power generation plants. Hydrogen and oxygen, if produced, could be transmitted to load centers via underground pipeline networks. Ocean-based systems would require facilities for transmitting electricity or piping or shipping hydrogen and oxygen to shore points from ocean-based system sites, about 25 to 160 miles offshore.

(3) Potential Economic Viability of System

(a) Capital Costs (1970 dollars)	Estimates of capital costs range from $168 to $400 per kw$_e$. Lower figure includes construction and engineering costs only. Upper figure is goal for capital cost of pilot plants. Principal factors governing capital cost of system include: area required for heat-exchangers; water volumes required; materials used to prevent fouling and corrosion; diameter, length,

(continued)

Criterion	Assessment
	stability and thermal insulation required for water pipes; size, pressure drop, thermal drop and working fluid of boilers, evaporators and condensers; turbine size and efficiency; parasitic pumping losses; types of energy transmission or transportation facilities used; size and efficiency of electrolysis facilities, if used; size of hydrogen storage subsystem, if used; etc.
(b) Operating Costs	Operating costs should be about the same as for a conventional electric power generation plant, in terms of cost per kw_e per year.
(c) Economies-of Scale	Preliminary studies indicate that the optimum plant-size for ocean-based system may be about 750 Mw_e.
(d) Estimated Price of System Products (1970 dollars)	About 10 to 15 mills/kwh by year 2000, assuming capital costs can be reduced to a level of about $200 to $400/$kw_e$ (rated capacity).

(4) Environmental Impacts of System

(a) Environmental Effects	Virtually pollution free. No residuals produced.
(b) Land-Use Requirements	Ocean-based systems would only require land for support facilities.
	Shore-based systems may require about 1 linear mile of coastline for each 1000 Mw_e plant, or about 1,000 miles of suitable coastline to satisfy about 10% of estimated U.S. energy requirements by the year 2020.
(c) Water Requirements	For electrolysis process, if used.
(d) Heat Balance	Preliminary studies show that systems that could supply 10% of U.S. energy needs by year 2020, if located in the Gulf Stream, might be expected to change water temperature, near the ocean surface, by less than 1°C (including mixing effects).
(e) Visual, Noise, Odor Effects	None.
(f) Ecological Problems	Studies should be made of the effects on the ecology of changing the temperature of the ocean near the surface (see 4d, above).
(g) Plant Failure	Plant failure would affect the energy supply provided by these systems but should have no other adverse effects.

(5) Sociological Acceptability of System

(a) Esthetics	Ocean-based systems would cause some esthetic impact because of the shore-based support facilities required.
	Shore-based systems would have an esthetic impact similar to that of any other power plant except that cooling towers, smokestacks or other tall structures would not be required. Moreover, if a large number of installations are required in

(continued)

Criterion	Assessment
	coastal regions of high tourist potential and unusual scenic value, these esthetic problems would probably be exacerbated.
(b) Noise, Odor, etc.	No problems anticipated.
(c) Safety	For ocean-based systems the safety of workers would be similar to that of other submarine-type operations.
	For shore-based systems the safety of workers would be similar to any operation requiring some offshore maintenance.
(d) Area Impact	May have some sociological impacts on Southern U.S. coastal areas caused by increased industrial activity, employment, housing, etc., in areas of operation.
(6) Institutional Constraints	
(a) Institutional Acceptability	Institutional acceptance will depend on whether it proves to be technically feasible and economically viable as indicated by Proof of Concept Experiments (POCE) and demonstration systems tests and evaluations.
(b) Implementability	This system will, as will alternative energy systems, cost hugh amounts of money to construct and operate. However, the returns on investments should be comparable to other energy systems, if the cost goals for the system can be met.
(c) Zoning	Zoning of large amounts of coastline will probably need to be altered to allow construction of shore-based systems or support facilities for ocean-based systems.
(d) Ownership	Systems would probably be owned by public utilities or, possibly, the Federal Government.
(e) Legal Problems	Legal problems may arise both from zoning regulations for coastlines, and international regulations governing the use of the oceans for ocean-based systems.

Problem Areas

Systems analysis of ocean thermal gradient systems indicated the possible problem areas shown in Table 1.4.

TABLE 1.4: PROBLEMS OF OCEAN THERMAL GRADIENT SYSTEMS

System Component	Technical	Economic	Environmental	Sociological	Institutional
Collection (ocean-based)	Heat-exchangers-size, efficiency, material Cold water pipe (size, material, stability, structure) Parasitic power (pumping) losses Turbine design and efficiency Underwater maintenance Open-cycle (flash evaporation of seawater) vs closed cycle (intermediate working fluid) Corrosion Biofouling Preserving thermal stratification of seawater Size and stability requirements of container hull, vulnerability to storms Optimum depth System trade offs Design for multiple use Plant sizing	Cost of heat exchangers Cost of material to meet constraints of corrosion and structural requirements Heavy capital investment of hull (floating platform) Number of plants required Cost of means to prevent biofouling Cost-effectiveness of afloat vs land-based system Cost-effective determination of mixes of multiple uses Cost-effective optimization of plant size	Possible leakage of working fluid Siting problems Effects of lowering water temperature Possible ecologic effects of numerous sites	Safety aspects Secondary impacts of environmental effects	Legality of plants outside continental limits Ownership and management of plants Financing Industrial practices (e.g., conservatism based on conventional considerations of efficiency Security of system (e.g., from sabotage, enemy attack, etc.)

(continued)

TABLE 1.4: (continued)

System Component	Technical	Economic	Environmental	Sociological	Institutional
Conversion (ocean-based)	Design of turbine and generator to meet distribution requirements Design of diffuser Sealing of turbine Turbine efficiency Materials study for marine environment Develop, test and evaluate mass production of conversion elements Design and operation of AC vs DC generators Time varying power production Efficiency of electrolysis units Electrolysis design DC to AC inverters	Life-cycle costs of conversion components Cost of mass-production of conversion components Cost-effectiveness of optimum materials for conversion components	Effects of temperature changes on biota Disposal of cooling water	Safety aspects of converters Public acceptability of hydrogen	Legal and international problems Intertie to public utilities Standards for products Security of converter system
Storage (ocean-based)	Choice of state of H_2 Requirements for storage at ocean-based site Means of providing storage at site	Life-cycle costs Economic viability Economics of ocean-based storage vs immediate piping	Impacts of possible leakage	Psychological aspects of H_2 storage Methods of improving storage safety	Institution problem of storing afloat National defense and security problems International problems associated with storage afloat

(continued)

TABLE 1.4: (continued)

System Component	Technical	Economic	Environmental	Sociological	Institutional
Transportation, Transmission and Distribution (ocean-based)	Transmission of electricity to shore points; Intertieing with public utility networks; Optimum means of transporting hydrogen from ocean-based sites	Cost-effectiveness of different media; Trade-off with direct use of electricity	Environmental impact of required shore-based facilities; Impact of leakage; Impact of pipelines or electrical transmission lines on coastlines	Safety problems; Public acceptability; Esthetic problems	Institutional acceptability; Security problems; International problems
Utilization (ocean-based)	Survey of loads; Methods of utilizing H_2, O_2, etc.; Fuel cells vs aphodid burners for converting H_2 to electricity; Improved electrode materials for fuel cells; Optimum designs for fuel cells, aphodid burners, etc.; Hydrogen for transportation units; Reversible electrolysis/fuel cell units	Survey of market potential for products; Availability of critical materials for fuel cells, etc.; Economics of using H_2 for households, industrial processes, etc.; Life-cycle costs of fuel cells, aphodid burners, etc.; Life-cycle costs of transportation units using H_2 vs fossil fuels	Impact of energy derived from system; Methods of reducing thermal pollution and utilizing waste heat	Public acceptability of H_2; Methods of improving safety of utilization systems; Survey of sociological considerations of OTGS; Plans for phasing in utilization of energy	Institutional barriers; Feasibility of colocating industries near shore points; Survey of building and zoning codes relative to fuel cells, aphodid burners and other end-use facilities; Regulations governing utilization of hydrogen derivatives; Price regulations and controls for products
Systems Integration (ocean-based)	Test, assembly and emplacement techniques; Suppression of marine growth; Operational requirements	Life-cycle cost of system; Cost of products to consumers; Economic viability of alternative types of systems	Impact of methods of marine growth suppression on biota; Short-term and long-term meteorological effects	Esthetic impacts of system; Social problems caused by siting; Land requirements; Availability of suitable land for shore facilities	Financing requirements; Institutional acceptability of system; Legal problems and international problems associated with offshore siting

(continued)

TABLE 1.4: (continued)

System Component	Technical	Economic	Environmental	Sociological	Institutional
	Trade-offs between open cycle and closed cycle systems Trade-offs between ocean-based and shore based systems Engineering specifications for selected systems Test facilities Optimum plant size Associated products and by-products Preventive maintenance	Trade-offs to optimize system efficiency Optimum product mix Optimum plant size Optimum spacing and siting of plants	Possible interference with Gulf Stream CO_2 balance in ocean Impact of leakage of working fluids	Public acceptability of system	National defense considerations and security considerations of OTGS (Ocean Thermal Gradient System)
Collection (heat-engine, including turbine) (shore-based)	Heat exchangers-size, efficiency, material Cold water pipe (size, material, stability, structure) Thermal leakage Parasitic power (pumping) losses Turbine design and efficiency Open cycle (flash evaporation of seawater) vs closed cycle (intermediate working fluid) Corrosion Biofouling Preserving thermal stratification of seawater	Cost of heat exchangers Cost of material to meet constraints of corrosion and pipe insulation Number of plants required Cost of means to prevent biofouling Cost-effectiveness of land-based system vs one afloat Cost-effective determination of mixes of multiple uses Cost-effective optimizing of plant size	Land use (difficulty of obtaining enough sites) Siting problems Effects of lowering water temperature Ecologic impact of numerous plants	Impact on land use Impact on socio-economic conditions of plant construction in region Other secondary impacts of environmental effects Esthetic considerations Safety aspects	Ownership and management of plants Financing Industrial practices

(continued)

TABLE 1.4: (continued)

System Component	Technical	Economic	Environmental	Sociological	Institutional
	System trade-offs Design for multiple use (e.g., mariculture, desalination, etc.) Plant sizing				
Conversion (shore-based)	Design of generators Efficiency loss in conversion Efficiency of electrolytic process	Low-cost production of electrolytic cells			
Storage			Same as for ocean-based system		
Transmission and Distribution (shore-based)		Cost-effectiveness of direct electricity vs hydrogen conversion	Environmental impact of transmission lines and other facilities		
Utilization		Same as for ocean-based system	Same as for ocean-based system		
Systems Integration		Same as for ocean-based system	Same as for ocean-based system		Optimum integration into power grid

Source: PB 231 142

OPEN CYCLE
OCEAN THERMAL GRADIENT POWER PLANT

The information in this chapter is excerpted from the following publications:

NASA TM X 70783
PB 228 066
PB 231 142
PB 238 571
PB 239 393

After the early proposal of D'Arsonval, the first important development of an ocean thermal powerplant system was carried out by Claude and his coworkers in the 1920s. Claude's system was based on the use of an open cycle, using the seawater as a working fluid, and included the construction of a prototype powerplant off the Cuban coast that delivered 22 Mw of power from the turbine. Further work using an open cycle was carried out by the French government during the 1940s and 1950s. There has been little work on open cycle ocean thermal powerplants in recent times.

Claude favored the open cycle on the basis that the high cost of the working fluid, the enormous size of the heat exchangers, and a large temperature drop across the boiler would make the closed system too costly. Based on Claude and Boucherot's calculations, 5,000 calories could be obtained per cubic meter of surface water if it were assumed that the warm water would be cooled 5°C during evaporation.

Furthermore, they concluded that 100,000 kg-m of energy could theoretically be extracted for every cubic meter by use of this open cycle. By assuming an overall efficiency of 75% so that 75,000 kg-m/m³ would be available, and by estimating that it would take no more than 30,000 kg-m/m³ of scavenge power to run the plant, they predicted a net power output of 45,000 kg-m/m³.

With an assumed flow rate of 1,000 m³/sec, a net power output of 400 Mw was predicted. These predictions demonstrate the feasibility of the open cycle system.

FIGURE 2.1: SCHEMATIC OF OPEN CYCLE VARIANT OF OCEAN THERMAL GRADIENT COLLECTION AND CONVERSION

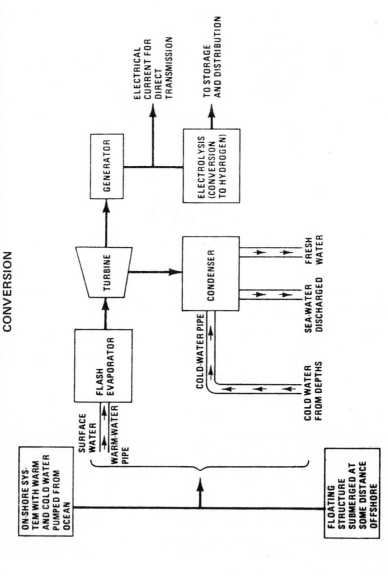

Source: PB 231 142

Figures 2.1 and 2.2 are schematics of an open cycle SSPP. This type of plant has its basis in the fact that one can lower the temperature water boils at by lowering the pressure upon it. Thus in Figure 2.1 the tepid ocean surface water will boil in the flash evaporator because the air has been evacuated from the chamber causing a partial vacuum in it. The steam generated thereby spins the turbine. The steam is then condensed via the cold deep ocean water. This condensed steam is pure (desalinated) water and constitutes one of the truly significant advantages of the open cycle over the closed cycle.

FIGURE 2.2: THE CLAUDE OCEAN THERMAL DIFFERENCE PROCESS

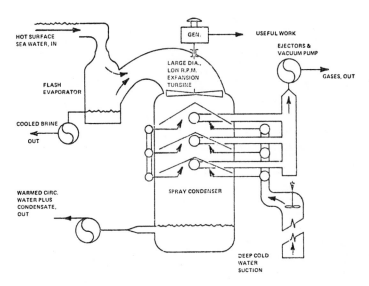

Source: NASA TM X 70783

One objection to the Claude cycle rests on the basis that the SSPP vapor pressure of only 25 to 30 mm Hg would require very large turbines. It is estimated that for a one million kilowatt plant a football field size nozzle throat would be required. It is perhaps unreasonable to build a one gigawatt SSPP with its 10,000 m^2 nozzle throat. A series of 10 Mw SSPPs may be preferable to one very large SSPP. Even though the turbines for the open cycle SSPP are relatively large the pressures and temperatures under which they must operate are relatively small. Thus these turbines can be made of lightweight materials such as plastics, fiber glass, composites, etc. It may be cheaper to build large plastic turbines than to build the relatively smaller, but stronger, closed cycle turbines.

The open cycle SSPP has the potential to produce up to 25% more power than the closed cycle SSPP. The reason for this is that there is no loss in temperature in the exchange of heat from the tepid surface water to the working fluid in the open cycle process. Also in the closed cycle SSPP the heat exchangers are the most expensive component of the plant. This cost is saved if the Claude process is used. Moreover, one need not contemplate the loss of volatile or hazardous working fluids in the open cycle.

Closed cycle systems are extremely sensitive to biofouling of the heat exchangers. This problem may severely limit the life of any closed cycle SSPP. Since boiling in the open cycle SSPP is primarily dependent upon the vacuum and not the temperature of heat transfer surfaces the biofouling constraint does not approach the criticality it does with the closed cycle case. If biofouling is the factor that most severely limits SSPP life, an open cycle SSPP would survive a great deal longer than a closed cycle plant. Even the Nitinol engine will be susceptible to biofouling and the formation of slime on the surface of the material which progressively cuts down on the transfer of heat from the tepid (and cold) water to the material will drastically cut down on its efficiency.

An equally convincing reason for building SSPPs with open cycle power generators is the additional use of cold water for fish farming. All SSPP designs make use of massive amounts of cold deep ocean water, therefore no one design possesses any significant advantage over any other with respect to mariculture considerations. However, the Claude cycle SSPP is the only system under consideration that produces fresh water as a by-product. The United Nations does not foresee any long term energy shortages, but predicts serious water shortages.

Professor Donald F. Othmer of Brooklyn Polytechnic and Mr. Oswald A. Roels of the Lamont-Doherty Geological Observatory have stated that if the cost of producing power is taken as 6 mills/kwh, fresh water costs are $1.38 per thousand gallons; or if power cost is taken as one cent/kwh, then fresh water costs $1.28 per thousand gallons. Thus a small SSPP built with off-the-shelf hardware could produce power at a cost of one cent/kwh.

The fact that many islands which are prime SSPP sites have shortages of water and many countries with prime SSPP sites are largely desert and the water desalinated by a Claude cycle SSPP may be more valuable than the power produced must be considered. Even at the site of the experimental closed cycle SSPP there are fresh water problems due to incessant pumping and use of the subsurface water. The aquifer level has been dangerously lowered and salt water intrusion into the aquifer is becoming a serious problem. A new fresh water source would be very welcome there.

In any case the value of the distilled and desalinated water produced in the SSPP by condensing the steam should be several times the value of the power produced. To summarize, the Claude (or open) cycle SSPP offers the following advantages:

   (a)   No loss in temperature due to the use of heat exchangers (increased
         power production)
   (b)   No need for expensive heat exchangers
   (c)   The production of desalinated and distilled water
   (d)   Relative immunity to biological fouling of surfaces
   (e)   Inexpensive turbines required
   (f)   Claude cycle SSPP technology will create the greatest demand for
         SSPPs by many nations due to (c) above.

It should be pointed out that the open cycle system had many technical problems. Recent work points out the disadvantages of an open Rankine cycle over the closed version; specifically, that the turbine design may be the key weakness of the open cycle version. However, the use of fluids other than water for large scale powerplants presents in itself a formidable problem.

While the past work of researchers in this area shows the overall feasibility of such a system, need for a more detailed technical updating of the system and its components exists. However, even though almost 50 years of engineering progress have elapsed since the time of Claude, many of the problems that they faced will not be overcome by a mere updating of his work. The following study to be presented will consist of a preliminary design study of the open cycle which was conducted in order to identify the specific technical problem areas in which future work must be concentrated. Based on this work, future research may circumvent these problems with alternative designs or to establish the feasibility of constructing certain components in extremely large sizes.

PRELIMINARY DESIGN INVESTIGATION

Overall System Design

The open cycle design presented here consists of four basic components:

(1) Evaporator—the design for this initial study is based on a spray evaporator chamber will give the largest possible surface area to enhance the heat transfer necessary for flashing;

(2) Turbine—the initial design will consider both axial flow and radial inflow single stage turbines;

(3) Deaerator—this component will be responsible for removing all non-condensable gases from the turbine exhaust;

(4) Condenser—the preliminary design for a condenser is a tube-in-shell heat exchanger. This configuration was selected so that the turbine exhaust vapor can be condensed into fresh water.

The system was designed for a gross turbine output of 100 Mw. It was felt that this would be a convenient size to determine the scavenge power requirements and sizes for a large plant. Figure 2.3 shows an overall schematic of the cycle and Figure 2.4 summarizes the results of the initial sizing of the systems components.

Design of the Spray Evaporator

For the design of the spray evaporator, it was assumed that the evaporator inlet was 30' below the ocean's surface to isolate it from the interference of the wave's motions. The water temperature at this depth was assumed to be 80°F. In order to achieve flashing, the pressure of the flash chambers had to be below 0.5069 psi (saturated vapor pressure at 80°F). A pressure of 0.36292 psi (saturated pressure at 70°F) was chosen as the design condition as it was assumed that a pressure below this might be too expensive to maintain. These assumptions set the inlet and outlet conditions for the evaporator as shown in Figure 2.5 and Table 2.1 (on p. 34).

FIGURE 2.3: OPEN CYCLE SCHEMATIC

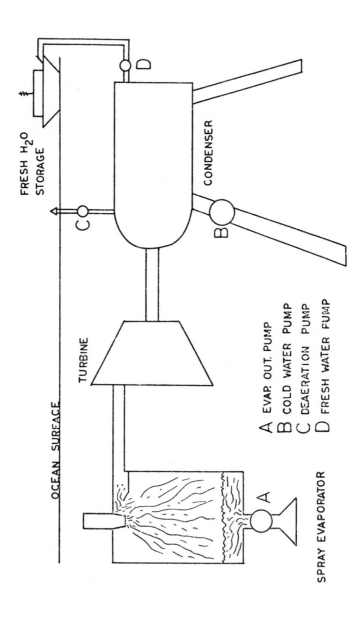

OCEAN SURFACE

FRESH H$_2$O STORAGE

CONDENSER

TURBINE

SPRAY EVAPORATOR

A EVAP. OUT. PUMP
B COLD WATER PUMP
C DEAERATION PUMP
D FRESH WATER PUMP

Source: PB 239 393

FIGURE 2.4: OPEN CYCLE POWERPLANT DESIGN

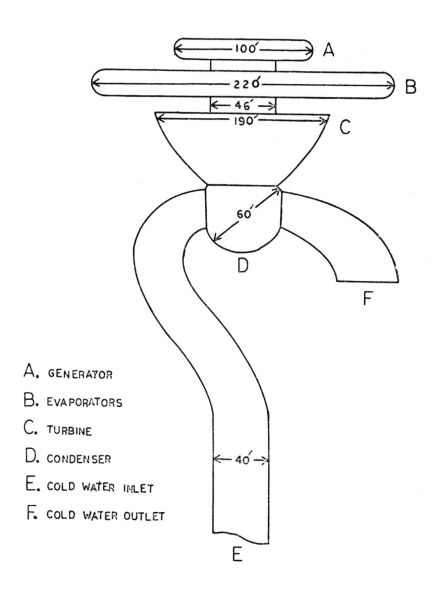

A. GENERATOR

B. EVAPORATORS

C. TURBINE

D. CONDENSER

E. COLD WATER INLET

F. COLD WATER OUTLET

Source: PB 239 393

TABLE 2.1: THERMODYNAMIC CONDITIONS OF SPRAY EVAPORATOR

State	Temperature, °F	Pressure, psia	Specific Volume, ft³/lb	Enthalpy, Btu/lb	Entropy, Btu/lbF
Seawater inlet	80	27.7	0.01607	48.11	0.0932
Vapor exit	70	0.36	868.4	1092.1	2.0645
Liquid exit	70	0.36	0.01605	38.05	0.0745

FIGURE 2.5: OPEN CYCLE EVAPORATOR SCHEMATIC

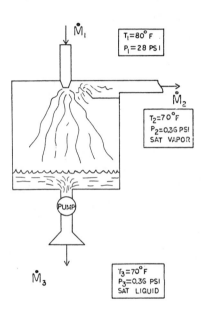

Source: PB 239 393

By assuming an adiabatic evaporator and that the system would come to thermodynamic equilibrium before the vapor and liquid leave the flash chamber, the ratio of the mass flow rate of vapor leaving the chamber to the mass flow rate of seawater entering the chamber was calculated from the conservation of mass and energy.

In order to get a better feel for the size of the spray evaporator, the number of nozzles necessary to handle the large flow of seawater was calculated. A Sprayco 18E full cone center jet nozzle made of Saran plastic was chosen as a typical large flow nozzle that was already in commercial use.

The pressure difference necessary to run the spray evaporator will initially be generated by vacuum pumps.' Once the system begins to operate the pressure difference across the evaporator's nozzles will be maintained by the condenser.

It will also be necessary to pump the unflashed water back into the ocean. The scavenge work necessary to operate the spray evaporator for a 100 Mw gross output plant was found to be 32.4 Mw.

Design of the Turbine

Since the pressure head across the turbine will be only 0.2 psi, there must be a mass flow rate of 1.73×10^5 lb/min in order to produce 100 Mw of gross output. Obviously, the turbine would have to be very large in order to handle this flow, but more importantly, it would have to operate with almost zero losses. Thus, this initial study was limited to single stage machines with no interstagial losses. The thermodynamic properties of the vapor at the inlet and outlet of the turbine are tabulated below.

State	T, °F	P, psi	V, ft³/lb	H, Btu/lb	X
Inlet	70	0.36	868.4	1092.1	1.00
Exit	50	0.178	1645.0	1048.1	0.966

Using values tabulated above, the enthalpy head across the turbine was found. Once the value for the specific speed and the specific diameter are determined, one can find the speed and size of the turbine. The following values for N_s (specific speed) and D_s (specific diameter) were determined for axial flow and radial inflow turbines by assuming maximum efficiency.

Turbine Type	N_s	D_s
Axial	120	1.2
Radial inflow	80	1.4

Appropriate calculations yielded the following results.

Turbine Type	Rotative Speed of the Turbine (N), rpm	Turbine Blade Diameter (D), ft
Axial	139.13	191.40
Radial inflow	92.76	223.30

Since the size of one turbine for this cycle is extremely large, it might be necessary to have several smaller turbines instead of one large one. By assuming that all turbines would operate under the same conditions and handle the same flow rate, it was possible to analyze the change in size and speed as the number of turbines was varied. Analysis showed that the relative size of the individual turbine diameters can be reduced greatly. For example, with five turbines, an axial turbine design would require a diameter of 86' and a radial inflow design would need a diameter of 100'.

Deaeration Losses

A problem in designing the condenser is that the turbine exhaust vapor contains a significant amount of noncondensable gases. These gases significantly lower the heat transfer coefficients on the vapor side. This problem must be overcome by deaerating the turbine exhaust. While the collection mechanism was left for future design considerations, it was possible to calculate the ideal deaeration work. The first step in determining the deaeration work was calculating the amount of gas within the turbine vapor. It was assumed that the seawater temperature at the evaporator's entrance was 80°F. At this point two additional assumptions were made.

(1) The seawater is at atmospheric pressure (even though the evaporator is 30' below the water's surface).

(2) The only noncondensables that are released are those contained by the seawater that actually flashes.

By having made the above assumptions and by utilizing the data contained in Table 2.2, it is possible to calculate the amount of noncondensables that would reach the deaerator, namely, 0.248 lb/sec gas.

TABLE 2.2: GAS CONTENT OF WATER, SATURATED AT ATMOSPHERIC PRESSURE

T, °F	Fresh Water, lb/million lb H_2O	Salt Water, lb/million lb H_2O
30	39	100
40	35	96
50	30	92
60	27	90
70	24	88
80	21	96
90	19	85

Source: PB 239 393

This flow of gas must be pumped from the condenser, where it is at 50°F and about 0.2 psi, to the ocean's surface where it will be expelled at 14.7 psi. By the use of the first law and the perfect gas law, the adiabatic pump work was found to be 0.08 Mw.

This value for deaeration losses seemed to be much too conservative. However, it was previously assumed that only the liquid that flashed to vapor would give up its noncondensable gases. The other extreme, and probably more realistic, would be to assume that all the liquid sprayed into the evaporator gave up its noncondensables. Since the inlet and outlet conditions of the deaerator for this case were the same as in the first case, the head did not change. However, the mass flow rate did. Therefore the deaeration pump work was found in this case to be 8.2 Mw.

Condenser Design

It was found that there were 2.88 x 10³ lb/sec of turbine exhaust at 50°F and
0.966 quality which the condenser must transform to the liquid state. Once
the total heat transfer rate for the condenser was calculated, the mass flow rate
of the cold ocean water needed to condense the turbine exhaust vapor was
found to be 5.95 x 10⁵ lb/sec.

It was assumed that the cooling water was brought to the condenser by a 40'
diameter tube. Since the condenser pressure was very low and it was very desirable to avoid any pressure loss, condensation was assumed to take place on
the outside of the tubes. The tube spacing was set so that the cross-sectional
area of the shell was twice the cross-sectional area of the tubes.

For a specified tube diameter, the mass flow rate in each condenser was calculated. Since the mass flow rate of the total tube bundle is the sum of the
individual tube mass flow rates, the number of tubes was found. The length
of the tube bundle was then determined.

A computer program was written which would calculate the geometry of the
condenser core by varying the tube diameter and the inside tube velocity. By
utilizing this program, a parametric study was conducted which showed the effect of varying the tube diameter and inside velocity.

While many condenser core geometries were shown to be possible, other factors
were taken into account in order to pick the best core. The pump work required
for the cooling water circulation is a direct function of the condenser core geometry. The losses that the cooling water circulation pump must overcome are a
combination of the loss due to the condenser core and the loss due to the inlet
tube.

The effect that inside tube velocity and tube diameter had on the condenser
pumping power was studied. The results of this study demonstrated that the
pump work goes down as the inside velocity and tube diameter go down.
Ideally, one would like to choose a small tube with a low inside velocity. However, it is also desirable to keep the tube lengths less than 50' for practicality.
For this reason a 1" nominal tube diameter was chosen (1.315" o.d., 1.049"
i.d.). In order to keep the number of tubes needed down and at the same
time keep the shell diameter down an inside flow rate of 10 ft/sec was chosen.
Once the tube diameter and the inside flow velocity were chosen, the core
geometry was set. The final condenser specifications are listed below.

Tube diameter	1" nominal
Shell diameter	48.57'
Number of tubes	154,276
Core length	40.76'
Inside velocity	10.0 ft/sec
Overall heat transfer coefficient	686.31 Btu/hr ft² °F
Total heat transfer area	2,165,000
Core pressure drop	6.80 psi
Cooling water pumping work	13.5 Mw

Sizing of the Cold Water Pump

Studies demonstrated that is is possible to design a cold water circulation pump that can be housed within the cold water inlet or outlet pipe.

Fresh Water Production

Once the turbine exhaust vapor is condensed in a tube-in-shell condenser, it can be collected for fresh water. There is sufficient mass flow in the 100 Mw gross output plant to produce 25 MGD of fresh water (about as much as a medium distillation plant). There are two alternatives for handling the fresh water once it has been produced.

(1) Pump it to a station on the water's surface which will be a combination storage facility and barge loading facility.

(2) Pipe the fresh water to the mainland via underwater pipelines.

It was assumed that the first few open cycle plants would probably pump the water to a surface station rather than incur the cost of underwater pipelines. The pumping requirements were calculated. The fresh water pump work was found to be 0.142.

Summary and Conclusions

To determine the feasibility of the open cycle system, a plant with a 100 Mw gross output was used. It was found that in order to get 100 Mw of gross output the turbine (assumed to be 75% efficient) would have to be approximately 200' in diameter and handle a flow of 2.88×10^3 lb/sec.

The condenser for the open cycle would consist of 154,276 1" nominal copper tubes that would be 40.76' in length in order to handle the cooling water flow of 5.95×10^5 lb/sec.

The evaporator would handle 1.82×10^7 lb/min of seawater since only 0.95% of the water that passes through the chamber would flash. The required scavenge work for this system is tabulated below.

	Mw
Evaporator exit pump	32.4
Cooling water circulation pump	13.5
Deaeration pump	8.2
Fresh water pump	0.2
Total ideal pumping requirements	54.3

Therefore for this plant there will be 45.7 Mw net output.

The results of this preliminary study show that many major modifications in the system design are required to make the open cycle system more attractive. Specifically, future design and analysis work should concentrate on the following components as well as preliminary system economic considerations.

(1) Evaporator—The preliminary design showed that it was possible to generate the vapor needed to operate the open cycle. However, it was assumed that the spray would come to equilibrium within the flashing chamber. It is therefore desirable to determine the time needed for the spray to come to equilibrium so a more realistic appraisal of the evaporator can be made. Once the equilibrium readjustment time has been computed, it will then be possible to find the size of the flash chamber.

If the spray chamber for the system turns out to be too large then it may be possible to design a many layer channel flow flash evaporator. This evaporator could be modeled on the basis of large desalinization plant practice.

(2) Turbine—The results of this preliminary study showed that a single turbine with a diameter of 190 to 223 feet would be needed to handle the mass flow rate of working fluid for a 100 Mw gross power output system. Obviously, such a large turbine size is impractical and this size problem must be overcome. Future work should continue on new turbine concepts and multiple, more efficient turbine designs should also be considered.

(3) Deaerator—In order to keep the condensation rate at a maximum level it is necessary to remove all noncondensables from the turbine exhaust. While the initial design attempted to determine how much scavenge work would be needed to deaerate the vapor, a much more in depth analysis is needed.

(4) Condenser—The preliminary design of the condenser used a shell and tube condenser. While such a design has the distinct advantage of producing fresh water, it is still necessary to determine whether other configurations may be more economically advantageous. Of particular interest are the many types of direct contact condensers.

FEASIBILITY STUDY OF A 100 MEGAWATT PLANT

An open Rankine cycle ocean thermal powerplant and its key components was designed for a gross turbine output of 100 Mw. This size was selected as representative of a reasonably large powerplant for determining the parasitic power losses. An overall schematic diagram of the proposed cycle configuration is found in Figure 2.6, and Figure 2.7 presents a scale layout. The open cycle design consists of four basic components:

(1) Turbine—The turbine is a most important component in the open cycle design since it will set the flow rates for the other components. Two turbine designs are to be considered. The first is a single stage radial inflow turbine. It was felt that one large turbine would facilitate the handling of the large flow of vapor with minimal losses. For this design, the vapor will enter from evaporator exits which surround the turbine. Inside the turbine, the vapor will naturally flow inward toward the center of the turbine where it will leave to enter the condenser as one stream. The other turbine design is a conventional axial flow design.

(2) Condenser—A tube-in-shell configuration was selected so that the turbine exhaust vapor can be condensed into fresh water which will be transported to the mainland. The value of fresh water is quite obvious as the demand for it grows daily.

FIGURE 2.6: OPEN CYCLE SCHEMATIC

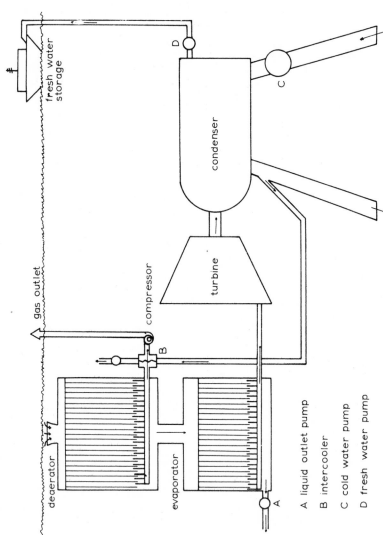

fresh water
storage

gas outlet

condenser

compressor

turbine

deaerator

evaporator

A liquid outlet pump

B intercooler

C cold water pump

D fresh water pump

Source: PB 238 571

FIGURE 2.7: OPEN CYCLE POWERPLANT DESIGN

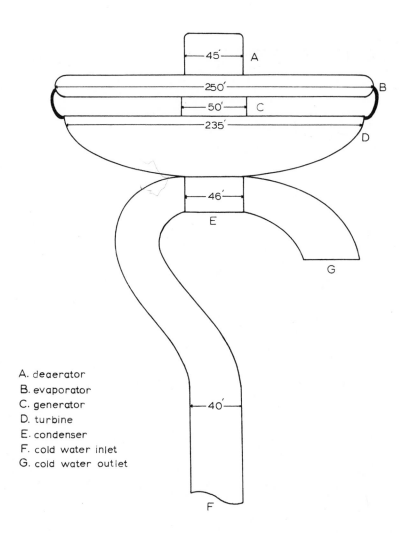

A. deaerator
B. evaporator
C. generator
D. turbine
E. condenser
F. cold water inlet
G. cold water outlet

Source: PB 238 571

An integral part of this design is the cooling water circulation pump which will draw 40°F ocean water up from 1,800 feet below the ocean's surface and through the condenser.

(3) Evaporator—Two types of evaporator designs were investigated in detail, a spray evaporator system and a falling film design. However, the need for a deaerator in the system meant the spray evaporator design had to be eliminated. Initially it was felt that the deaeration process for this system would be placed between the turbine outlet and the condenser inlet. This left a large pressure difference between the ocean and the evaporator which would supply the driving force needed to operate the spray evaporator.

In view of the tremendous flow rate of water the system must handle, rapid evaporation of the ocean water is essential. By atomizing the ocean water as it enters the evaporator, an extremely large droplet surface area could be achieved so that evaporation would take place very rapidly.

However, it is impractical to deaerate the turbine exhaust vapor since it is almost impossible practically to separate the gas from the vapor. The ideal time to deaerate the ocean water is just before it enters the evaporator. This would be done by lowering the pressure of the liquid stream which would drive the air from the liquid.

In order effectively to deaerate the water, it is necessary to drop the pressure to such a level that it will no longer be possible to operate a spray evaporator. A falling film evaporator would enhance the heat transfer in the evaporator by dividing the incoming water into many thin sheets. The surface area to volume ratio for this technique is not as high as that for a spray; however, the pressure differential is not required since the water naturally flows down the evaporator due to the force of gravity.

(4) Deaerator—Two different deaeration schemes were investigated. The first achieves the desired degassing by lowering the pressure of the water in stages. Using a staged system makes it possible to utilize conventional compressors to evacuate the air from the different deaeration chambers. However, a much simpler method would be to have the water pressure lowered in only one stage. This could be accomplished by evacuating the deaerator with a jet-ejector powered by the cooling water from the condenser exit. The only parasitic power required by this system would be that used to pump the ejector exhaust overboard.

Design of the Turbine

As pointed out in the previous study, the turbine must be quite large. Listed below are the properties of the inlet vapor and the saturated liquid and saturated vapor properties at the turbine exit.

State	T, °F	P, psia	V, ft³/lb	H, Btu/lb	S, Btu/lb °F	X
Inlet	70	0.36	868.4	1092.1	2.0642	1.00
Exit (sat liquid)	50	0.178	0.01602	18.054	0.0361	0.00
Exit (sat vapor)	50	0.178	1704.0	1083.4	2.1263	1.00

As stated above a radial inflow turbine would best fit the open cycle configuration. The vapor would flow out of ducts leading from the bottoms of the falling film tiers to the perimeter of the turbine where the ducts would become nozzles so that full admission could be achieved. The vapor would travel towards the center of the turbine where it could easily be piped to the condenser in one large tube.

In order to design the radial inflow turbine for maximum efficiency (and hence minimize the flow, a direct measure of component size, through the entire system) a conventional turbine design procedure was carried out.

The final radial inflow turbine design based on the appropriate calculations was:

Turbine efficiency	93%
Specific diameter	1.525
Specific speed	70
Mass flow rate	1.5237×10^5 lb/min
Outlet quality	97.31%
Volumetric flow rate	4.2109×10^6 ft^3/sec
Turbine diameter	235.42 ft
Turbine speed	80.12 rpm
ϵ-ratio of tip diameter to root diameter	1.55
Root diameter	151.88 ft
Turbine blade height	83.54 ft

To insure that it will be possible to fabricate turbine blades from available materials, a simplified stress analysis was performed. The results of the analysis showed the stress to be well within the strength limitations of many engineering materials.

Based on a report on open cycle ocean thermal difference turbines it was found that the optimum efficiency for a full admissions turbine would require a specific speed of 100 and a specific diameter of 1.3. It was also shown that for a 100 Mw plant the turbine diameter would be 2,500" (208.3'), the blade height would be 500" (41.6'), and the rotative speed of the turbine would be 112 rpm.

The study indicated that the optimum turbine size for a 75% efficient turbine would be around 0.6 Mw. This would mean there would have to be around 180 individual turbines to produce the desired 100 Mw.

While conventional turbine design techniques have been employed here, it is quite easy to see that a whole new approach is needed if the open cycle is to become a reality. The idea of having many turbines might seem practical from a present turbine economics standpoint; however, the pressure head across the turbine is small enough so that losses associated with splitting the system into many small fragments might render the scheme impractical.

Falling Film Evaporator Design

As was shown above, 1.5237×10^5 lb/min of saturated vapor at 70°F are required by the turbine in order to produce 100 gross Mw of electricity.

As previously stated, the ocean water must be deaerated before it enters the evaporator. This will be accomplished by reducing the pressure of the liquid stream so its pressure will be 0.522 psia as it enters the evaporator. In order to achieve the desired evaporation, the operating pressure of the evaporator must be below 0.5069 psia (the saturated vapor pressure at 80°F). For design purposes, a pressure of 0.3629 psia (the saturated vapor pressure at 70°F) was chosen as the evaporator operation pressure. It was felt that any pressure above this would make the evaporator excessively large and any pressure below this would make the turbine and/or condenser too large. These assumptions set the inlet and outlet conditions for the evaporator as tabulated below and shown on Figure 2.8.

State	Temperature, °F	Pressure, psia	Specific Volume, ft³/lb	Enthalpy, Btu/lb	Entropy, Btu/lb °F
Evaporator inlet	80	0.522	0.01607	48.11	0.0932
Vapor exit	70	0.36	868.4	1092.1	2.0645
Liquid exit	70	0.36	0.01607	38.05	0.0745

FIGURE 2.8: FALLING FILM EVAPORATOR SCHEMATIC

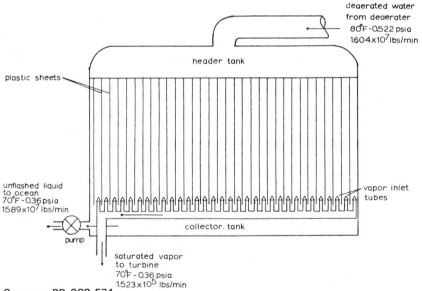

Source: PB 238 571

Since the vapor flow rate requirement for the turbine is 1.5237 x 10⁵ lb/min, the evaporator inlet flow rate must be 1.604 x 10⁷ lb/min in order to satisfy thermodynamic equilibrium.

Due to the large heat transfer area required to handle an inlet flow rate of

1.604 x 10^7 lb/min of ocean water, a compact evaporator design was a necessity. After considering many designs, a system of concentric plastic sheets was chosen for the evaporator since it would be an inexpensive, yet compact design. As Figure 2.8 shows, the sheets would be fed by header tanks which are supplied by the deaerator with water free of air. There will be slits in the bottom of the header tank to let a film of water (at a specified thickness) flow down both sides of each plastic sheet.

The sheets will be 2" apart with vapor inlets spaced at regular inlets near the bottom of the sheets. It is assumed that the vapor will come off the film and flow down the sheet towards the vapor inlet tubes at approximately the same speed as the film of liquid. (This assumption implies zero interfacial shear stress.) Once the vapor enters the inlet tubes, it will be piped toward the turbine. The liquid film will flow into a collection tank below the set of concentric sheets, from which it can be pumped back into the ocean. Because of their geometry, it is quite convenient to stack sets of these concentric sheets on top of one another.

It should be pointed out that the diameter of the innermost ring was set at fifty feet. This was done so that the deaerator and generator could comfortably be housed inside of the evaporator above the turbine. In this way ocean water will enter the deaerator (which will also be a set of concentric plastic sheets) and then flow to the header tanks of the evaporator free of air. All this would be accomplished by gravity (obviously the bottom of the deaerator would be above the evaporator) and a system of liquid pumps would finally discharge the unflashed water overboard. The greatest parasitic loss for this system is that work required to pump the unflashed water back into the ocean.

Since the horizontal width (or combined circumference) of the concentric plastic sheets was found earlier in the analysis, the length of the sheets was determined by dividing the total area by the horizontal width. Two feet were added to the length to account for the header tank and the collector tank.

A series of geometries were calculated for film thickness of from 0.007" to 0.013". As was previously mentioned, the innermost sheet had a diameter of 50'. The outer sheet diameter varied from 75' to 500' in 25' increments. The number of tiers of evaporator units was found by summing up all the horizontal film width available (to the given outer diameter) and dividing the horizontal film width required to accommodate the inflow.

In order to find a final design, the outer diameter of the evaporator was chosen so that it would be very convenient to feed the turbine nozzles with the vapor collected at the end of the evaporators as shown in Figure 2.6. Since the radial turbine blade diameter was determined to be 235', an outer evaporator diameter of 250' was chosen. A 10' limit was placed on the length of the plastic evaporator sheet since any length greater than this might lead to severe sheet buckling or failure. The final design is summarized below.

Inner diameter	50'
Outer diameter	250'
Height	15.52'
Film thickness	0.010"

Tiers	2
Reynolds number	293.3
Heat transfer coefficient	505.6 Btu/hr ft^2 °F
Required area	41,299,877.6 ft^2
Height of sheet	5.76'
Distance between sheets	2"
Required pump power	27.61 Mw

This evaporator design is not necessarily the optimum design for this system. Other considerations such as the shell that houses the evaporator to protect it from the sea may greatly alter the concept of an optimum evaporator design. However, the design presented here will form a reasonable starting point for a final optimum design iteration process.

Deaeration of the Ocean Water

The presence of noncondensable gases has an adverse effect upon the heat transfer processes that occur within a condensing heat exchanger. For this reason the open cycle must have a deaerator to remove the noncondensable gases from the incoming ocean water. The required deaeration of 1.604×10^7 lb/min of ocean water is clearly one of the disadvantages of the open cycle. However, as will be shown, the parasitic work for this proposed system is small.

At 80°F the gas content of fresh water (saturated at atmospheric pressure) is 21 ppm by weight as compared with 96 ppm for seawater. The higher gas content of the ocean is due to the presence of carbonate ($CO_3^=$) and bicarbonate (HCO_3^-) in solution. The following set of equations shows the equilibrium of the carbonate system in seawater.

$$CO_2 \text{ (gas)} \rightleftharpoons CO_2 \text{ (solution)}$$

$$CO_2 \text{ (solution)} + H_2O \rightleftharpoons H_2CO_3$$

$$H_2CO_3 \rightleftharpoons HCO_3^- + H^+$$

$$HCO_3^- \rightleftharpoons CO_3^= + H^+$$

One can see by inspecting the above equations, that as the partial pressure of the CO_2 above the solution is lowered carbonate ions ($CO_3^=$) must form bicarbonate ions (HCO_3^-) which in turn from carbon dioxide. However, for the open cycle design the presence of the carbonate system can be ignored since the first order reaction rate constant at 25°C for the reaction $HCO_3^- \longrightarrow CO_2 + OH^-$ is 2×10^{-4} sec^{-1}. In other words, since the reaction is so slow, the ocean water will have been flushed overboard long before any appreciable amount of CO_2 could have been formed. Thus, for the purposes of this analysis, it will be assumed that seawater has the gas content of fresh water.

The basic method of deaeration is to lower the pressure of the air surrounding the water, which in essence means lowering the concentration of the surrounding gases. This disruption of the equilibrium between gases in solution and the surrounding gases can only result in gas leaving the water.

The method of deaeration examined for this system was one which utilized compressors to lower the deaeration chamber pressure.

Calculations showed that this scheme would be totally impractical and it was suggested that this problem could be alleviated by staging the deaeration process. Instead of having one compressor, it would be possible to have a series of five compressors evacuating five deaeration chambers. The compressors are arranged in series so that the compressors not only have to handle the air-water vapor flow from their own stage but also the flow from the previous compressors.

While this series of five compressors was a vast improvement over a single compressor, the system was improved further by utilizing a portion of the cooling water from the condenser exit to cool the gas mixture to 60°F before it was compressed.

The design results showed that lowering the temperature significantly reduces the water vapor flow since most of it will condense out. This results in a much lower parasitic work requirement for the deaeration system. Intercooling, however, requires a pump to move the cold water from the condenser through the intercoolers and finally back to the ocean. In order to find the real savings of the intercooled system, the parasitic work for this pump had to be determined and was found to be 170.12 kw (0.17 Mw).

It was concluded that even though the intercooling system required 0.170 Mw of additional parasitic work to power a cold water pump, it would be a savings of 1.75 Mw since the compressor work would only be 0.44 Mw as compared with 2.36 Mw for a system without intercooling.

The design chosen for the five deaerator stages was very similar to the falling film evaporator design. This scheme, which gives a high surface area to volume ratio, will keep the dimensions of the deaerator stages to a minimum. The basic assumption underlying this design was that the mechanism of degassing would be similar to that of evaporation. This assumption meant that the heat transfer correlations previously used to size the falling film evaporator could be employed to size the deaerator stages. It was also assumed that the mixture of air and water vapor would act homogeneously with an appropriate mean molecular weight.

In the evaporator design, a cylinder with a 50' diameter was left empty within the center where the deaerator and generator would be housed. From that central location it would be very convenient for the deaerator to supply the header tanks of both evaporators with degassed water. The design of the deaerators was also chosen to be a set of concentric sheets having the exact same setup for gas collection, header feed tanks, etc. The main difference was that the innermost sheet would have a diameter of 6" and all other sheets will be 6" apart.

Figure 2.9 shows the final design for the deaerator-compressor system. The inlet header tank was assumed to be 2' high. Since each tank in between two stages will be a combination collector and header tank, they will be assumed to be 4' high. The final collector tank (from which the two evaporation tiers will be fed) was also assumed to be 4' high. A summary of the final intercooled deaerator-compressor design follows.

Number of deaerator-compressor stages	5
Deaerator diameter	45'
Total deaerator height	32'
Intercooling mass flow rate	54,983.53 lb/min
Total compressor work	0.44 Mw
Intercooling water pump work	0.17 Mw
Total deaeration parasitic work loss	0.61 Mw

FIGURE 2.9: FINAL STAGES DEAERATOR DESIGN (with intercooling)

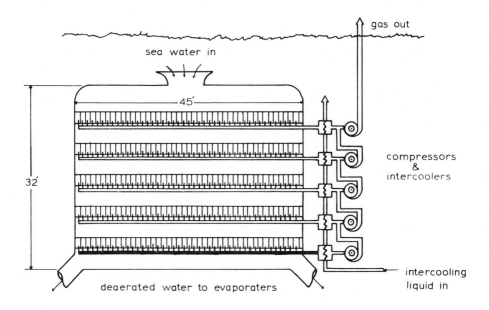

Source: PB 238 571

Condenser Design

As was shown above, there would be 1.5237 x 10^5 lb/min of turbine exhaust at 50°F and 0.9731 quality which the condenser must transform to the liquid state. The thermodynamic end points of the condenser are tabulated below.

State	Temp, °F	Pressure, psia	X	H, Btu/lb
Condenser inlet	50	0.178	0.9731	1054.78
Condenser outlet	50	0.178	0.0	18.1

It was assumed that the cooling water was brought to the condenser by a 40' diameter tube. Since the condenser pressure was very low and it was desirable to avoid any pressure loss, condensation was assumed to take place on the outside of the tubes. Following conventional design, the tube spacing was set so that the cross-sectional area of the shell was twice the cross-sectional area of the tubes.

While there are many possible condenser core geometries, other factors must be taken into account in order to pick the best core. The pump work required for the cooling water circulation is a direct function of the condenser core geometry. The losses that the cooling water circulation pump must overcome are a combination of the loss due to the condenser core and the loss due to the inlet tube.

By studying the effect that inside tube velocity and tube diameter had on the pump work, it was shown that the pump work decreases as the inside velocity and tube diameter decreases. So, ideally a small tube with a low inside velocity should be chosen. However, it is also desirable to keep the tube lengths less than 50' for practicality. In order to keep the number of tubes and pump work required down and at the same time keep the shell diameter down, a 1" nominal tube diameter having an inside flow rate of 10 ft/sec was chosen. Once the tube diameter and the inside flow velocity were found, the core geometry was set. The final condenser specifications are listed below.

Tube diameter	1" nominal
Shell diameter	46.0'
Number of tubes	137,500
Core length	39.1'
Inside velocity	10.0 ft/sec
Overall heat transfer coefficient	692.89 Btu/hr ft^2 °F
Total heat transfer area	1,954,052.94 ft^2
Core pressure drop	7.08 psi
Cooling water pump work	12.55 Mw

Sizing of the Cold Water Pump

While the cold water pump will necessarily be very large to handle the tremendous cooling water flow rate it is technically feasible. Calculations indicated that it is possible to design a cold water circulation pump that can be housed within the cold water inlet or outlet pipe. One design based on these calculations is summarized below.

Type	Propeller pump
Head	17.45'
Flow rate	3.78 x 10^6 gpm
Outer rotor diameter	38.75'
Inner rotor diameter	19.38'
Shape number	500
Rotative speed	37.6 rpm

Fresh Water Production

Once the turbine exhaust vapor has been condensed in a tube-in-shell condenser, it can be collected for fresh water. There will be sufficient vapor mass flow rate in the 100 Mw gross output plant to produce 25 mgd of fresh water (about as much as a medium size distillation plant). There are two alternatives for handling the fresh water once it has been produced.

(1) Pump the water to a station on the water's surface which will be a combination storage and barge loading facility.

(2) Pipe the water to the mainland via underwater pipelines.

It was assumed that the first few open cycle plants would probably pump the water to a surface station rather than incur the cost of underwater pipelines. The fresh water pump work was calculated to be 119.3 kw.

Summary and Conclusions

To determine the feasibility of the open cycle system, a plant with a 100 Mw gross output was studied. It was found that in order to produce this output, a radial inflow turbine designed for maximum efficiency would have turbine blades with a 235.42' diameter. The efficiency for this turbine was found to be 93% with a water rate of 1.52×10^5 lb/min.

The evaporator and deaerator would have to handle 1.60×10^7 lb/min of sea-water to produce the 1.52×10^5 lb/min of vapor required by the turbine since only 0.95% of the water that would pass through the chamber would flash. The deaerator would consist of five stages of concentric sheet-falling films, each with an inner sheet diameter of 6" and and outer sheet diameter of 45'.

The height of the entire deaerator would be 32'. The deaerated gas would be cooled from 80° to 60°F in the intercoolers before being ejected to the atmosphere by a series of compressors. A single stage evaporator was based on a two tiered system of concentric sheet-falling films. Its inner sheet diameter would be 50' and its outer sheet diameter would be 250'. The entire evaporator would be approximately 16' in height.

The condenser for the open cycle would be used to condense the turbine exhaust into 25 mgd of fresh water to be pumped to a storage facility on the ocean's surface. This condenser would consist of 137,500 1" tubes that would be 39.1' in length to handle the cooling water flow of 3.16×10^7 lb/min needed to condense 1.52×10^5 lb/min of turbine exhaust.

The required parasitic losses for this system are tabulated below.

Evaporator exit pump	27.61 Mw
Cooling water circulation pump	12.55 Mw
Deaerator parasitic work	0.61 Mw
Fresh water pump	0.12 Mw
Total parasitic pump work	40.89 Mw

Therefore, this plant will produce 59.11 Mw net electrical output.

From this study, it can be concluded that the open cycle itself is economically infeasible as an electrical power production system, due to the extravagant cost of state-of-the-art turbines. For example, a 75% efficient turbine will cost at least $2,100/kw. It can sefely be assumed that even the radial inflow turbine design discussed here will be prohibitively expensive. However, the open cycle has great potential for producing only fresh water. The system presented here (minus the turbine) would be ideal for the production of fresh water.

CARNEGIE-MELLON UNIVERSITY DESIGN

The information in this chapter is excerpted from the following publications:

PB 228 068
PB 228 069
PB 235 469

A solar sea power plant (SSPP) generates electricity from the latent work potential of the temperature difference at the same location between the warm surface
waters of the tropical oceans and the cold layers of water below the thermocline
at depths of about 2,000 to 4,000 feet, due to deep currents from the earth's
poles. The technology of the plant's operation is based upon the thermodynamics cycles of heat engines. That is, a working fluid with a high vapor pressure
at the available temperature passes through a heat exchanger in which heat is
extracted from the upper layer of sea water warmed by the tropical sun and
pumped through the exchanger.

At a high-pressure vapor the working fluid passes through a turbine generator
which produces electric power. At the turbine exit the low-pressure vapor passes
through a condenser in which heat is rejected to the cold lower-layer seawater
pumped through the condenser. The working fluid vapor liquefies and is pumped
from the condenser back to the heat exchanger, completing the closed cycle of
the plant. Because of the small Carnot and conversion efficiencies involved, the
SSPP operation involves the handling of large volumes of water. (It has been
noted that the total amount of water to be handled may be less than the amount
required to produce the same amount of power in a hydroelectric plant.)

The primary output of the plant is electric power for local consumption or transmission, or for conversion into a storable and transportable fuel, such as hydrogen or ammonia. Depending on conditions at an individual site the SSPP may
be viewed further as the nucleus of a three-commodity economic system where
in addition to electric power output, (1) desalination equipment is added to
produce fresh water and (2) facilities are also included to support a mariculture
industry. That is, tropical regions are often in short supply of fresh water. The

production of fresh (distilled) water as a by-product of the SSPP output could become a valuable additional source of revenue through sales to the coastal regions adjoining a plant site. Similarly, the deep seawater of the ocean contains large quantities of nutrients which could be used to support a mariculture farming cycle feeding algae which feed shellfish which feed crustaceans, in a series of shallow pools. The wastes in turn support seaweed growth of value. Notwithstanding the potential economic appeal of this three-commodity view, the basic goal for the SSPP is the low-cost production and high-capacity provision of energy output.

Beyond the technical and engineering considerations associated with the plant design, construction and installation, it is apparent that the implementation of an energy system based on SSPP generation involves a variety of issues, ranging from conventional economics to the laws of the sea.

In the design topology for a SSPP, it is desirable that the design should not be dependent upon ocean currents to power the heat exchanger or to insure sufficient warm water supply. Such SSPP may therefore be located wherever the ocean thermal structure is favorable.

Contamination of the input water to the boiler by the output water is avoided by taking advantage of the naturally occurring density stratification, at least for plants which do not exceed 700,000 kilowatts. High-density power generation without moisture, and the consequent power expenditure in its removal, is obtained by falling film evaporation on vertical tubes. Appreciable pressure drop in the large banks of vertical tubes in the evaporator and condenser is avoided by proper manifolding.

The operating principle adopted by CMU design of SSPP is based upon the concept advanced by D'Arsonval in 1881. According to this concept one circumvents the enormous turbines required when one uses the seawater itself as the working medium. This is done by using an intermediate working fluid, such as ammonia, which has a sizable vapor pressure at ambient temperature.

Because of the low Carnot efficiency of such a plant, tremendous quantities of warm seawater must be passed through the heat exchanger which serves as a boiler for the ammonia liquid, and also tremendous quantities of cold seawater must be passed through the heat exchanger which serves as the condenser for the ammonia vapor.

As an example, in a 200,000-kw plant, the water supply requires a flow of 8 ft/sec through a 60-ft-diameter pipe. Because of this tremendous water flow, extreme precautions must be taken to avoid unnecessary losses, such as losses accompanying flow in curved pipes. This requirement of low flow loss has led to the design layout shown in Figure 3.1. The warm water, brought from above through a vertical pipe, is forced down through vertical tubes on the outside of which is a falling film of ammonia liquid. Collectively these tubes form the evaporator, commonly mislabelled as a boiler.

The cold seawater, brought from below also through a vertical pipe, is forced upwards through vertical tubes upon the outside of which ammonia vapor is condensing. Collectively, these tubes are called the condenser. Optimization studies show that, because the cold water pipe is much longer than the warm water pipe, the warm water flow is considerably stronger than the cold water flow. As a consequence, when the warm downward flowing water which exits from the evaporators meets with the cold upward flowing water which exits from the condenser, the two flows combine to form a downward flowing jet which eventually vanishes below the thermocline.

Losses must also be minimized in the working fluid flow as well as in the warm and cold water flow. In minimizing the water flow losses we have arranged for the water to flow vertically within the tubes. Geometry now requires that the ammonia vapor flow transversely to the tubes in both the evaporator and the condenser. But because of the very large number of tubes, in the order of one million for a 200,000-kw plant, the evaporator and condenser tube banks are several hundred tubes thick.

FIGURE 3.1: SKETCH OF A SOLAR SEA POWER PLANT

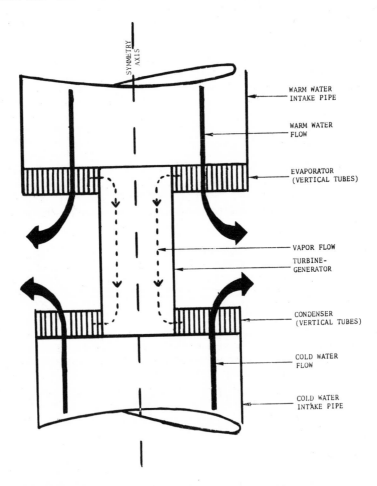

Source: PB 228 068

Conventionally designed tube banks of such thickness would have vapor pressure losses of orders of magnitude higher than we can tolerate. In designing the tube banks so as to minimize the maximum pressure drop within the banks, keeping the overall packing density constant, it has been found that if the bank is N tubes deep, the final design has a maximum pressure drop of only ½N times the pressure drop in a conventionally designed bank.

OVERVIEW OF THE DESIGN

An overview of the proposed layout for a SSPP is presented in Figure 3.1.

Water Circulation System

Warm water from above is forced through a vertical tube evaporator, thereby vaporizing the working fluid liquid (WFL), falling down as a film on the outside of vertical tubes. The central cylindrical section of the evaporator serves as a vapor collector. This vapor then flows downward through an axial gas turbine, which in turn powers a generator. After leaving the gas turbine the working fluid vapor (WFV) continues downward into the central cylindrical section of the vertical tube condenser, and condenses upon the surface of vertical tubes cooled by deep water forced up from below. The condensed working fluid (WFL) is then pressurized and pumped back into the evaporator.

The warm water flowing downward meets the cold water flowing upward. Since the warm water current is stronger than the cold water current, the resulting stream forms a downward conical jet which disappears below the thermocline boundary. The disappearance of this jet below the thermocline insures that the exhaust from neither the evaporator nor from the condenser contaminates the water in the warm water intake. The large distance from the cold water intake insures that this intake likewise remains free of mixing. An outstanding feature of this design is its perfect symmetry about a vertical axis. This symmetry provides many advantages:

(1) All changes in direction of flow, of both the water and the working fluid, come about through a change in pressure distribution of the fluid itself, rather than through a change in shape of the boundaries. All curved boundaries introduce particularly high frictional losses. These losses are all avoided by having the interaction of the two jets provide for the appropriate directional velocity changes.

(2) The absence of channels to provide changes in direction results in a reduction not only in power losses, discussed above, but also in the capital cost of the plant.

(3) The cylindrical system insures a relatively simple stress system in all structured parts, thereby simplifying the design for structural rigidity, and thereby reducing the probability of structural failure by design error.

Working Fluid Circulating System

Another outstanding feature of this design is that the details of the evaporator and condenser are essentially the same. It is recognized that evaporation and condensation are essentially reverse processes of each other. Figure 3.2 illustrates the reversibility of the elementary processes of evaporation and of condensation. The practical importance of such a reversible design is that a technical advance in either the evaporator or condenser has a high probability of being applicable to the other.

As indicated in Figure 3.1, the evaporator chamber contains a multitude of vertical tubes. This vapor must flow through a deep bank of these tubes into the vacant inner cylindrical region. The maximum pressure drop accompanying such flow must remain very small compared to the total available pressure drop across the turbine.

FIGURE 3.2: COMPARISON OF EVAPORATION AND CONDENSATION

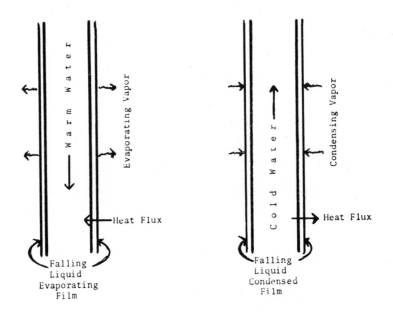

Source: PB 228 068

A 200,000-kw SSPP might contain up to 2,000,000 evaporator tubes. A uniform distribution of such tubes is indicated in the upper part of Figure 3.3. Such a uniform distribution would give an unacceptably high pressure drop unless the gap width (w) becomes comparable to the tube diameter. An overall high packing ratio can, however, be maintained while still keeping the pressure drop small, provided an appropriate manifold system is adopted. A possible manifold system is illustrated in the lower part of Figure 3.3. It is found that such manifolding reduces the maximum pressure drop by a factor of 1/N, where N, the number of circular rows, is of the order of several hundred.

FIGURE 3.3: MANIFOLDING SYSTEM

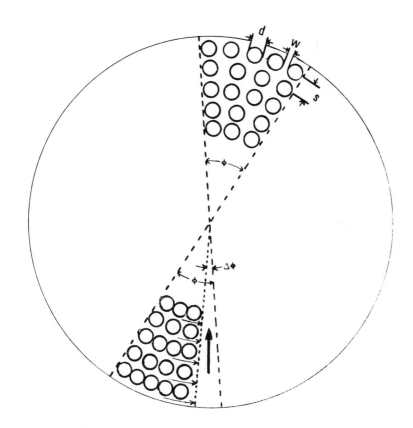

Source: PB 228 068

In a manifolded evaporator the vapor coming from a typical tube flows first in an azimuthal direction to the nearest radial channel, then along this radial channel to the central collector region. Precisely the same manifolding system is available to the condenser. Here the vapor flows from the central region out radially along a main channel, and then turns into a branch azimuthal channel to reach a particular condensing tube.

The liquid condensed on the condenser tubes must somehow be returned to the evaporator tubes, after passing through a pressurizer. The path the liquid must follow is essentially the reverse of the vapor path. The liquid collects in the bottom of the bottom of the condensers, and flows by gravity to the bottom of the cylindrical space. In this flow, the liquid encounters the same type of impedance from the bank of the condenser tubes that it previously encountered as a vapor. It will therefore follow the same channels as does a vapor, but in reverse. The liquid first flows azimuthly into the main radial channels, and then directly into the central cylindrical empty space. There, it is pumped into a header on the top portion of the evaporator. It now flows along the main radial channels, into the side azimuthal channels, and finally is ejected through holes in the header onto the top of the evaporator tubes.

SEAWATER INLET-OUTLET HYDRODYNAMICS

Topology of SSPP

The seawater system analysis deals primarily with an investigation of the field of seawater flow external to the solar sea power plant (SSPP). The fundamental details of the SSPP geometry and the typical sea environment are shown in Figure 3.4. The SSPP is represented as a vertical cylinder of length L, and diameter D. The ratio of L/D is expected to be of the order of 40:1. Schematically the warm water for the boiler heat exchanger enters at the top of the cylinder near the sea surface, and the cold water for the condenser heat exchanger enters at the bottom of the cylinder. The locations of the discharge outlets are at points to be determined along the length of the cylinder.

The simplest mathematical description of the SSPP is a vertical line with two point sinks and two point sources used to represent the intakes and outlets. The sources represent points where seawater with initial momentum and possible buoyancy (if the density match is not perfect) is discharged into the thermally (and density) stratified ocean. The sinks are the two points where seawater is selectively withdrawn from the stratified environment into the SSPP.

The configuration of the plant is chosen to assure that there is no short circuiting between the sources and sinks. The complete elimination of short circuiting would not be possible theoretically if the surrounding seawater were uniform in temperature and density. However, since the seawater is stratified thermally, it is theoretically possible to withdraw water of a particular temperature in a manner similar to withdrawing an individual sheet of paper from a stack of papers. By choice of direction of momentum flux and relative density match, it is possible to return seawater to the stratified ocean and assure that the effluents do not mix with the inflow at the inlets.

Relation of Plant Size and Seawater Flow

The work obtained from the temperature difference in the ocean depends on the volumetric flow of seawater through the plant. The heat flux that must be exchanged can be calculated from the product of the mass flow of water, its temperature difference, and the efficiency of the plant. For example, assume seawater weighs 64.0 pounds force per cubic foot, its specific heat is 1 Btu per pound per °F, the efficiency is ϵ %. A numerical example calculated when $\Delta T = 4.5°F$, and $\epsilon = 2.5$ percent, shows that a flow of 132 cubic feet per second (cfs) or about 60,000 gpm is needed to generate 1,000 kw.

Two-Layer System: Model for Warm Water Intake

The warm water inlet can be modeled as an axisymmetric flow to a sink in a two-layer system. The warm water intake is located in the region where the thermal gradient is nearly zero assuming that the region above the thermocline is constant

FIGURE 3.4: TOPOLOGY OF A SOLAR SEA POWER PLANT

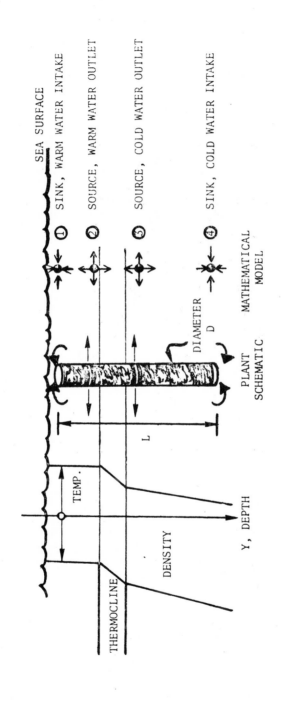

Source: PB 228 068

in both temperature and density. Figure 3.5 shows the region schematically. The seawater below the interface (thermocline) is slightly colder and therefore more dense than the seawater above the interface. At low rates of inflow, only the seawater above the interface is withdrawn.

FIGURE 3.5: CRITICAL FLOW FROM TWO-LAYER SYSTEM—WARM WATER INTAKE

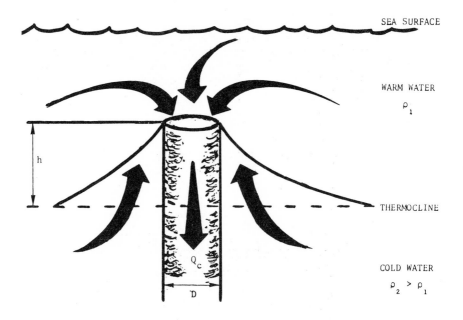

Source: PB 228 068

However, at higher flow rates both regions will cntribute to the flow at the intake. Investigations have been made to determine the critical value of the withdrawal rate in terms of other known quantitites, such as distance from the interface and density difference across the interface. For a critical withdrawal rate of 100,000 cfs, the above conditions limit the plant capacity to about 0.76 million kw where the upper layer is at 25°C and the lower layer is at 20°C if it nominally takes a flow of 132 cfs per 1,000 kw capacity.

Intake Water Structure

Precautions to insure against the intake of debris and aquatic life must be provided at the SSPP. The SSPP may typically have a flow of 60,000 to 100,000 gallons per minute per megawatt capacity. Material is kept out of the cooling water pipe by physical screening at the entrance. Trash racks halt the passage of the largest debris and fine screens provide the final obstacle for the flow. The selection of screen mesh size is generally based on removal of trash that could clog the condenser tubes. An engineering rule of thumb for selecting mesh size is that the clear openings in the screen should be limited to about one-half the diameter of the condenser tubes. While protection against clogging of the condenser tubes is of primary importance, it is necessary to limit the smallness of the screen openings to maintain a relatively low head loss across the inlet screens.

DESIGN FEATURES

Boiler Technology Considerations

Nucleate vs Spray Boiling: Pool boiling at the low super heat (ΔT_{SH}) available to us has a notorious low heat transfer coefficient. As illustrated in Figure 3.6, at the lowest ΔT_{SH}), heat is carried to the top of the pool by natural convection, and vaporization then takes place at the liquid-vapor interface. At higher ΔT_{SH}'s, vapor bubbles form on the metal-liquid interface.

FIGURE 3.6: GENERAL CHARACTERISTICS OF POOL BOILING

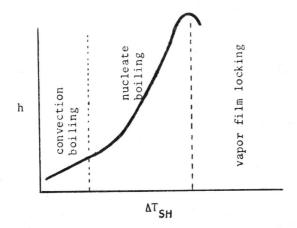

Source: PB 228 069

As illustrated in Figure 3.7, the rapidly expanding bubble has a thin layer of liquid which separates the main part of the bubble from the wall. This thin layer offers only a small impedance to thermal flux from the metal wall surface to the liquid-vapor interface. Under ordinary conditions a ΔT_{SH} of over 10°F is required for nucleate boiling. However, in this case there is not 10°F to spare as superheat.

A major problem exists in transferring the vapor entrapped in the bubbles to a pure vapor phase, even to a vapor phase with a low moisture content. It is therefore pertinent to consider an alternative method of vaporization in which

FIGURE 3.7: EXPLODING BUBBLE

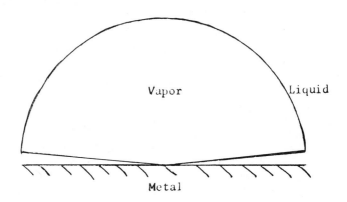

Source: PB 228 069

the vapor passes directly into the vapor phase. In such a method a thin film of fluid falls down the evaporating heat exchange surface.

Working Medium or Water in Intake Tube: Either the working medium (e.g., ammonia) or the seawater could move through the tubes. Thus in Figure 3.8a, ammonia moves up in tubes; the seawater has a cross horizontal flow. In Figure 3.8b, ammonia moves up between the tubes; the seawater flows horizontally within the tubes. In both cases, the direction of flow is determined by the requirement that the liquid ammonia flows upward towards its vapor phase.

FIGURE 3.8: POSSIBLE BOILER TOPOLOGIES

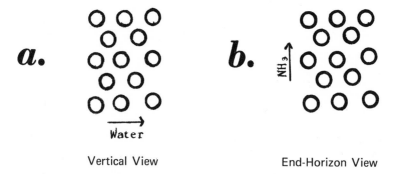

Source: PB 228 069

It is possible that either arrangement in Figure 3.8 could be acceptable. It is also possible that some asymmetry in the role of water and of ammonia may render one arrangement of overwhelming advantage. It is therefore pertinent to look at the asymmetric roles of water and of ammonia. Each gram of water which passes through the boiler gives up about two calories of heat. Each gram of ammonia which passes through the boiler absorbs about 320 calories. Therefore, about 160 times as much mass of water flows through the boiler as mass of ammonia.

This asymmetry in mass flux far outweighs the 50% additional volume flux of ammonia vapor leaving the boiler over the volume flux of water through the boiler. The high mass flux of water through the boiler requires that particular care be taken in avoiding all unnecessary pumping losses.

Seawater Pressure≤Working Medium Pressure: Once the topology of Figure 3.8b is chosen, the requirement of elastic stability requires that the hydrostatic pressure of the water exceed that of the working fluid. The tubes would be subject to elastic collapse if their external pressure were to exceed their internal pressure. Specifically, when this pressure difference exceeds the critical value $\Delta P_c = (1/48)(t/R)^3 E$ where t = wall thickness, R = radius, E = Young's modulus of elasticity, an initial, perfectly round cylinder will fail as indicated in the following Figure 3.9.

FIGURE 3.9: COLLAPSE OF PIPE

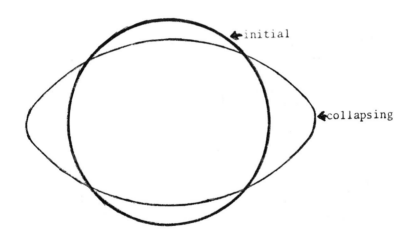

Source: PB 228 069

Other considerations may also place restrictions on the tube wall thickness t. Thus, the tubes may be called upon to furnish some structured rigidity to the boiler as a whole. Again, corrosion rates impose a minimum wall thickness.

If ammonia is the working medium, with a vapor pressure of 160 psi at 26°C, the boiler must be at a depth of at least 320 feet in order that the ocean pressure exceed 160 psi.

Topology of Vapor Drainage Within the Boiler: The depth of a pool is ultimately limited by the change in vapor pressure with depth. In the particular case of ammonia, at seven feet below the surface of the pool, the boiling temperature is 0.5°C higher than at the surface. The temperature gradient which drives the heat flux from the seawater into the ammonia is therefore 0.5°C less seven feet below the surface than at the surface.

Since this 0.5°C is a substantial fraction of the available 5°C for this gradient, it is doubtful that an optimization analysis will allow a depth greater than seven feet. Since, however, the total depth of our boiler must be much greater than seven feet, the conclusion that the boiler must be layered as indicated in Figure 3.10 is reached.

FIGURE 3.10: ILLUSTRATION OF LAYERED STRUCTURE OF BOILER

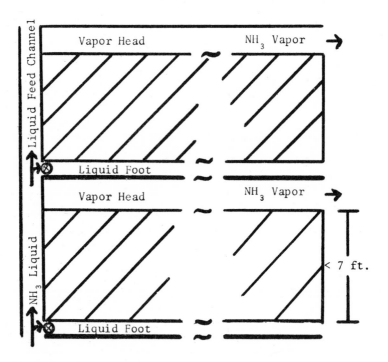

⊗ valve which introduces NH_3 at a predetermined pressure corresponding to a head of not more than seven feet. Thatched areas are pool boilers containing straight tubes running horizontally into plant of paper, as indicated in Figure 3.8b. The warm seawater flows through these tubes.

Source: PB 228 069

Dynamics of Pool Boiling: It is desirable that all the heat exchange surfaces in the boiler proper, indicated by thatching in Figure 3.10, operate in such a manner that the thermal impedance on the boiling side is low compared to that on the seawater side. Only then can a low heat exchange area per unit power output be maintained. It is further desired to keep that fraction of the boiler volume occupied by the ammonia liquid and vapor comparable to, certainly not much greater than, the fraction of volume occupied by the seawater. Only then can excessive cost of containing structure be avoided.

Finally, the ammonia vapor should emerge into the vapor head essentially free of liquid ammonia entrainment. Only then can the cost and power losses needed for entrained liquid separation be avoided. Such separation is necessary to avoid excessive erosion of the turbine blades.

The basic problems of boilers may all be traced to the extremely low thermal conductivity of liquids. Obtaining heat flux rates across heat exchange surfaces of more than 10,000 Btu/hr ft^2 is desired, the heat flux from the boiling side of the heat exchange surface to the liquid-vapor interface being driven by a temperature drop of not more than 1°F. But if a temperature drop of 1°F is to drive a heat flux of 10,000 Btu/hr ft^2 across a slab of still water (or ammonia), this slab must be less than 0.003 inch thick. In contrast, the distance from the heat exchange surfaces to the top of the pool may be of the order of 3 feet.

One method of accelerating heat flux is by convection. However, because the liquid velocity must be zero at a solid interface, convective heat transport cannot operate all the way to the surface. The general behavior of convective heat transport near a solid surface is illustrated in Figure 3.11a. Not only does the convective heat transport go to zero at the solid liquid interface, but also both its first and second derivative. One can thus truly speak of a laminar layer across which heat transfer must be by conduction.

Another method of minimizing the role of liquid thermal conduction is by nucleate boiling. Here the only molecular thermal conductance is across a thin film separating an exploding bubble from the solid wall. Buoyant forces then raise the bubbles to the top of the pool. A determination must be made, however, as to whether the bubbles rise fast enough to make it feasible to use a dense packing of heat exchange tubes, thereby avoiding excessive containment costs.

Suitable calculations give a vertical flux of heat (F_H) <2.6 cal/sec cm^2 or <34,000 Btu/hr ft^2, for water at 1 atm using the following values appropriate to water at 100°C, 1 atm pressure of density of vapor (ρ_V) = 0.60 x 10^3 g/cm^3 and heat of vaporization (H_V) = 574 cal/g. A considerably higher maximum heat flux in the case of ammonia in a SSPP boiler is anticipated for here the pressure, and hence the vapor density, is considerably higher. The values appropriate for boiling at 20°C are ρ_V = 0.0067 g/cm^3, H_V = 284 cal/g which leads to an F_H <14.2 cal/sec cm^2 or <186,000 Btu/hr ft^2 for ammonia at 20°C.

This maximum vertical heat flux given for an ammonia pool is far too small to give a compact boiler. The horizontal heat flux in the warm water in Figure 3.11 is approximately 480 cal/sec cm^2, corresponding to a 2°C change in temperature and a flow velocity of 8 ft/sec. In comparison, the maximum heat flux in the ammonia is only 14 cal/sec cm^2. The ammonia would therefore have to occupy about 30 times the volume occupied by water.

FIGURE 3.11: BEHAVIOR OF CONVECTIVE HEAT TRANSPORT NEAR A SOLID SURFACE

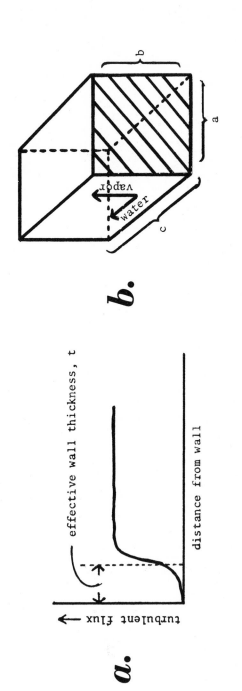

a.

Convective Heat Transport

b.

Comparison of Vapor Flux Emerging from
Boiler with Water Flux through Boiler

Forced Convection Boiling: In boiling a fluid, there are a variety of choices as to which variables are to be independent, which variables are to be dependent. In pool boiling, an area A is submerged in a pool. At least in controlled experiments, this area is maintained at a fixed temperature T. The mass flux G into the pool is adjusted to such a value to maintain a constant level in the pool. G is thus a dependent variable.

The roles of A and G are frequently reversed. Suppose there is an indefinitely long vertical tube maintained at a constant temperature T, above the boiling temperature. Fluid at a given mass flow rate G is injected into the bottom of the tube. This injected fluid is considered to be at the boiling temprature. For a sufficiently small mass flow rate G, and small superheat, boiling will take place by the rise of individual bubbles. A sharp boundary will exist between the fluid and vapor phase. The height of this boundary, and hence A, will increase as G is increased.

Hence, A is now clearly a dependent variable. As G increases, the heat flux at the top of the liquid column becomes too great to be sustained by the rise of individual bubbles. A foam phase will therefore develop. At a fixed G, the higher in the tube one goes, the larger is the mass fraction of vapor, i.e., the higher is the quality of the steam. In practice, the tube would be terminated at that height where the steam quality had reached a satisfactory level. The following table summarizes the difference between pool and forced convective boiling.

	Independent Variables	Dependent Variables
Pool boiling	$\Delta T, A$	$G(\Delta T, A)$
Forced convective boiling	$\Delta T, G$	$A(\Delta T, G, quality)$

The vertical tube discussed above represents only a simple example of forced convective boiling. The tube could be horizontal. Other possible configurations include boiling outside heating coils. The essential distinction between pool and forced convective boiling lies in which variables are treated as independent, and which as dependent, rather than in the geometry.

There thus appears to be no basic physical principle which would prevent the eventual design of a boiler with the following operating characteristics:

Cross flow as indicated in Figure 3.8b.

Volume occupied by working medium comparable to, and not much larger than, the volume occupied by the water.

Essentially zero liquid entrainment in vapor.

Because of the slow rise of bubbles in fluids and because of the low power available for separating moisture from vapor, film evaporation is preferred to pool boiling.

Cold Water Pipe vs Ammonia Pipe

Cold water from at least 2,000-ft depth must be used to condense the ammonia vapor. A major decision must be made in locating the condenser. The condenser may be put near the boiler, in which case the ammonia vapor has only a short distance to travel. However, the cold water must then be brought up from a depth of at least 2,000 feet. Alternately, the condenser may be placed where the cold water is located, and thereby avoid having to transport the cold water up 2,000 feet.

In this second case, the ammonia vapor will have to be brought down the 2,000 feet from the boiler, and the condensed ammonia pumped up to the boiler. Since the mass flow rate of the ammonia is only $1/150$ that of the cold water, it would appear that the second alternative would be preferable.

Thermodynamics of Cycle with Boiler and Condenser at Different Depths: Ignoring all irreversible effects, the net power output will be the same in the above-discussed two topologies. This equivalence is demonstrated in Figure 3.12, which illustrates the choice of the condenser being at the same depth as that of the cold water.

FIGURE 3.12: SSPP SYSTEM WITH CONDENSER AT DEEP WATER

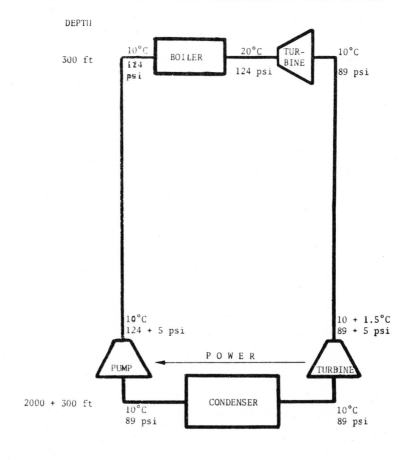

Source: PB 228 069

After leaving the boiler, the ammonia vapor passes through a turbine which extracts the same power as if the condenser were at the same depth as the boiler. In flowing adiabatically down the 2,000-foot pipe, the vapor pressure is increased 5 psi by the weight of the vapor above. Simultaneously, the temperature rises by 1.5°C.

The pressure and temperature now return to their original values when the vapor passes through a second turbine. The power from this turbine is now used to pump the 10°C condensed liquid ammonia back up to the boiler.

Ammonia Vapor Pipe: The long ammonia vapor pipe has one disadvantage over the long cold water pipe: the internal vapor pressure is far less than the outer hydrostatic pressure. Whether the wall strength required to withstand this pressure imbalance will render the ammonia vapor pipe impractical must be determined. Appropriate calculations indicate that the ammonia vapor pipe has the advantage over the cold water pipe of a smaller pipe radius, by a factor of 0.44, but the disadvantage of a higher wall rigidity, by a factor of 170. It appears that the disadvantage far outweighs the advantage.

Cold Water Intake Pipe

Wall Rigidity: A unique feature of SSPP's is the necessity of bringing tremendous quantities of cold water from the ocean depths in order to dissipate the heat given off by the condensers. Resistance to the mechanical forces imposed by the ocean environment must therefore be taken into consideration. Failure of structures caused by mechanical forces takes place by two distinct mechanisms: (1) failure by plastic yielding, (2) failure by elastic instability, i.e., by buckling.

Design for adequate strength to resist plastic yielding is relatively simple. One has merely to insure that the stress everywhere is within safe limits. Design to avoid elastic instability is more subtle. The designer must be sensitized to the various types of instability to which his structure is susceptible.

Wall rigidity can be increased more economically by forming a compound wall of two concentric cylinders, than by merely increasing the thickness of a single wall. Thus in Figure 3.13 two wall structures of the same rigidity are illustrated one a single 1" thick wall, the other two $\frac{1}{8}$" walls spaced $1\frac{5}{8}$" apart. The two walls must of course be appropriately separated. It has been suggested that a thin corrugated sheet, with an appropriate plastic filler be used. A filler which would render the pipe neutrally buoyant might even be found.

The double-walled structure has a second advantage over the single-walled structure. Steel in thick sections is much more susceptible to brittle failure than is steel in thin sections, particularly at temperatures as low as 5°C. In fact, plain carbon steel in sections $\frac{1}{8}$" thick will be essentially trouble-free from brittle fracture.

A third advantage of the double-walled structure is its relative ease of fabrication. In particular, thick-walled sections are weldable only if expensive special alloy steels are used. Again, for $\frac{1}{8}$" walls, plain low-carbon steels are weldable. The difference in weldability is traceable to the poor microstructures which arise from slow cooling of the welded zone. A $\frac{1}{8}$" plate weld can be quenched 64 times as rapidly as can a 1" plate weld.

FIGURE 3.13: COMPARISON OF TWO WALL STRUCTURES HAVING THE
SAME RIGIDITY (6.6 x 10^{12} ERGS)

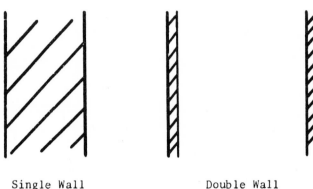

Single Wall   Double Wall

Source: PB 228 069

Failure by Collapse: Wall collapse represents one possible type of failure. Hydrostatic pressure considerations point out that the pump must either be put at the
bottom of the cold water pipe, or pay the penalty of an increased pipe rigidity.
Motion of the water outside the cold water pipe can also cause instability collapse.

Failure by Local Buckling: Under the ideal flow conditions represented in Figure 3.14, a pipe would experience no net force in an external current. Under
actual flow conditions, a net force does, of course, arise. The ocean current
force together with anchoring at the top and bottom, causes a moment across
the midsection of the pipe. This moment in turn induces a compressive force
on the upstream side of the pipe. This compressive force, if sufficiently large,
will cause local buckling.

For the wall of Figure 3.13, with a length equal to 1,000 meters, a velocity
≤170 cm/sec or ≤5.6 ft/sec is calculated. Such a velocity is well below that
anticipated within most parts of the Caribbean.

Failure by Snaking: Uniform one-dimensional flow of water in a pipe is intrinsi-
cally unstable. Such instability must be suppressed by the rigidity of the pipe
with respect to bending. Thus, visualizing a small dent in a pipe with zero rigidity, as in Figure 3.15a, the inertial forces acting on the pipe are indicated by
arrows.

In the absence of a stabilizing rigidity, the pipe will further deform, and will
pass through the exaggerated dent of Figure 3.15b (the same instability that causes
a river to wind snake-like back and forth in a gentle valley). Suitable calculations
lead to the conclusion that only very thin walls are required to suppress snaking
instability. The conclusion reached concerning the cold water pipe as a result of

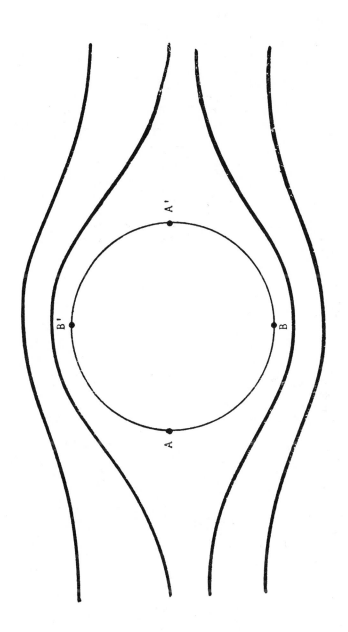

FIGURE 3.14: COLLAPSE BY OCEAN CURRENT

these considerations is that elastic instability, the mode of failure most difficult to guard against, may be controlled by two concentric ⅛" pipes spaced 2" apart, with an appropriate filler.

FIGURE 3.15: ILLUSTRATION OF SNAKING INSTABILITY

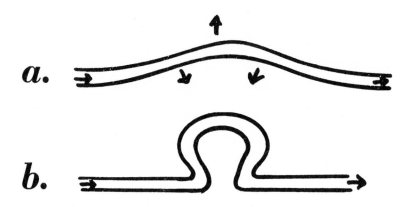

Source: PB 228 069

Water Pump

The cold water pump must supply power to:

(a) Raise the velocity of the intake water from zero to its final value v, tentatively taken as 8 ft/sec.

(b) Overcome the pressure drop within the cold water pipe, the length tentatively taken as 3,000 ft, the radius R as 30 feet.

(c) Overcome the pressure drop within the condenser.

(d) Raise the cold dense water up through warmer, and hence not so dense, water outside the pipe.

A SSPP with a cold water intake described in (a) and (b) above will develop a gross power of 185,000 kw when the intake water is warmed by 2°C. Using the values given in (a) and (b) above, the power needed to increase water velocity (P_1) is 2,000 kilowatts. The power needed to overcome pressure drop in the cold water pipe (P_2) is 720 kilowatts. The power needed to overcome pressure drop in the condenser (P_3) is 8,000 kilowatts.

Work for Pumping Cold Water: Since density decreases with a rise in temperature, work will be required. Figure 3.16 depicts a pipe which extends down to the 5°C level, filled with water from this depth. In order to evaluate the excess of pressure outside the pipe over that inside the pipe at any level above the 5° level the temperature profile T(z) must be known.

FIGURE 3.16: HYDRAULIC HEAD FOR COLD WATER PIPE

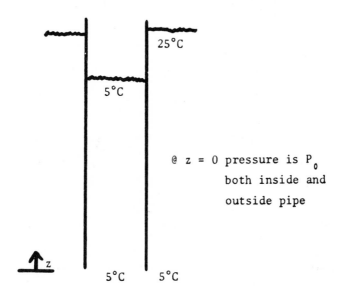

25°C

5°C

@ z = 0 pressure is P_0
both inside and
outside pipe

z

5°C   5°C

Source: PB 228 069

Typically this profile has the general form of Figure 3.17, at least in tropical
and near-tropical oceans. The upper layer has an essentially constant temperature. The bottom of the isothermal layer abuts a region in which the temperature drops suddenly.

FIGURE 3.17: SCHEMATIC TEMPERATURE PROFILE

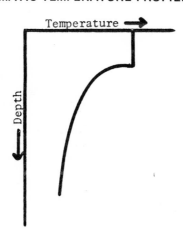

Temperature

Depth

Source: PB 228 069

FIGURE 3.18: TYPICAL TEMPERATURE DISTRIBUTION IN THERMOCLINE

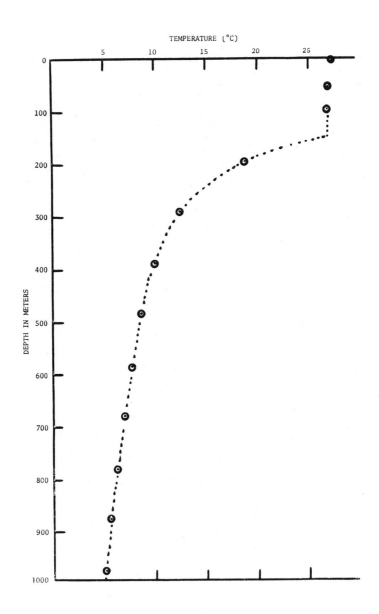

Source: PB 228 069

The observations of an actual temperature profile are reproduced as Figure 3.18. This profile was taken on the meridian passing through the Virgin Islands at that point where the surface current is the strongest. Transformation of this temperature profile into a density profile may be obtained from a table of the density of $35°/00$ salinity water as a function of temperature which is given in Figure 3.19.

FIGURE 3.19: DENSITY VARIATION WITH TEMPERATURE

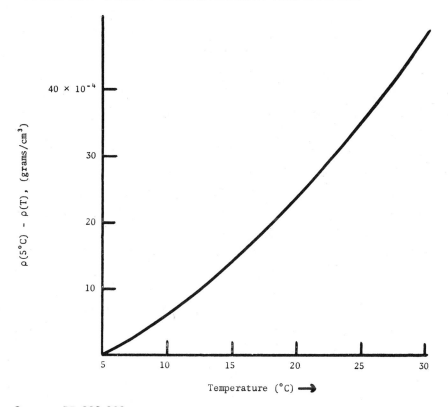

Source: PB 228 069

Combining Figures 3.18 and 3.19 gives Figure 3.20 which shows the density of the cold water in the pipe in excess of the density outside the pipe. The excess pressure, ΔP, outside the tubes over that inside of the tubes is plotted in Figure 3.21.

At the bottom of the isothermal layer, ΔP is only 50 g/cm^2, corresponding to a head of 0.5 meter. It then increases to 130 g/cm^2 at the surface of the ocean. The condenser of a SSPP will be located below the thermocline. Hence, ΔP will actually be less than 0.5 meter. Since the gross power output corresponds to a head of 25 meters (for a $2°C$ increase in cooling water temperature), less than 2% of the gross power would be expended in raising the cold water from the $5°C$ depth to the SSPP. This work does not, of course, include frictional losses.

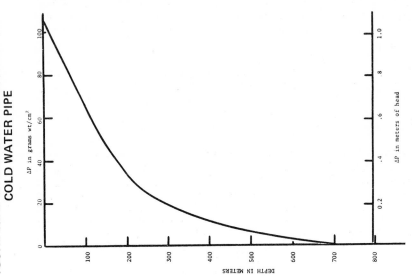

FIGURE 3.21: HYDRAULIC PRESSURE INSIDE COLD WATER PIPE

Source: PB 228 069

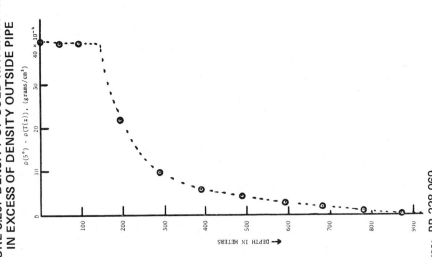

FIGURE 3.20: DENSITY OF COLD WATER IN PIPE IN EXCESS OF DENSITY OUTSIDE PIPE

Source: PB 228 069

Location of Pump: As mentioned above, the pressure inside the pipe should be greater than the pressure outside. It is therefore necessary that a pump at the bottom of the cold water pipe supply the power P_1 and P_2. Two distinct advantages would accrue from having one big pump at the bottom of the pipe which would also supply P_3 rather than two separate pumps, one at the bottom and one at the top of the cold water pipe.

First, the deeper the pump, and hence the greater the ambient pressure, the less severe is cavitation. Thus at a depth of 1,000 meters, the tip velocity of the pump blades would have to exceed 450 ft/sec to induce cavitation. Second, for a given length blade, blade rigidity becomes less of a problem the greater the design load.

A possible disadvantage of having one big pump at the bottom of the cold water pipe is that the resulting pressure within the pipe may require an excessive wall thickness. Calculating the wall thickness necessary to contain the internal pressure generated by having the total power $P_1 + P_2 + P_3$ supplied by one bottom pump indicates that for an allowed stress of 15,000 psi a steel wall thickness of 0.14", which is close to the ⅛" inner wall assumed previously would be required. Therefore, it is believed that internal pressure will not prevent having the total power $P_1 + P_2 + P_3$ concentrated within one pump at the bottom of the cold water pipe.

Pump Design: A feasible, and certainly the simplest, pump design is that of an axial flow propeller type. Such a design is based upon the well-known properties of an aerofoil.

Design studies indicate an overall design as shown in Figure 3.22 with the following design characteristics: Aerofoil dimensions, 18 ft span and 16 ft chord; an aerofoil mean velocity of 64 ft/sec; and an aerofoil inclination, $(\phi) \cong 10°$. Such a cold water pump will supply the pressure drop to pump the cold water up through a 3,000-ft-long cold water pipe, as well as through the condenser tubing.

Gas Turbine

The study of the thermodynamics, design principles and velocity design of the turbine for the SSPP indicated that the available temperature drop of 10°C for the heat engine is sufficient to drive a single-stage turbine with 50% reaction in stator and in rotor, and with maximum gas velocities of 740 ft/sec.

Antifouling Cost

Any surface immersed in the ocean rapidly acquires a film of bacteria slime. Later, except for copper and for surfaces covered with an antifouling paint, macro sea life grows in the slime-covered surface. The slime, let alone the later growth, would ruin any initial high heat transfer rate.

When, however, the surface is exposed to seawater containing as little as 0.25 parts per million of chlorine, all growth is prevented, even the bacterial slime. Therefore, an electrolysis system should be established which chlorinates the seawater coming into the boiler, the chlorine coming from the naturally occurring salt within the seawater itself, and the electric power coming from the power generated by the sea plant itself. An arrangement must of course be

FIGURE 3.22: A FEASIBLE PUMP DESIGN

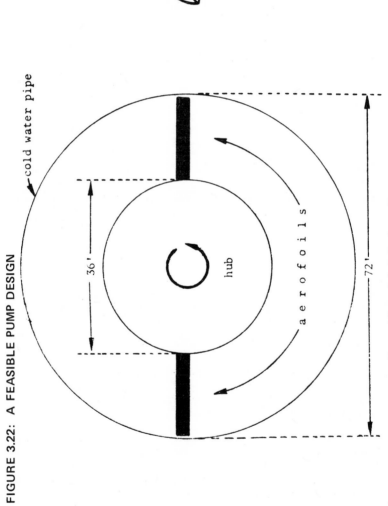

Pump Viewed Axially from Bottom

provided for the seawater surrounding the H_2 generating electrodes to bypass the condenser. In order for the proposed system to be economical, the power required for the chlorination must be small compared to the power generated by the sea plant. Calculations indicate that this is indeed the case.

Vertical Tube Heat Exchangers

Potential for Use of Fluted Surfaces in Evaporation and Condensation: The heat exchangers of a solar sea power plant are seriously constrained by (1) low over-all temperature differences across the heat transfer surfaces and (2) low pressure drop across the turbine. Consequently, heat transfer coefficients must be relatively high and essentially dry vapor must be produced with a minimum of auxiliary equipment.

The process of nucleate boiling can lead to appreciable entrainment of liquid in the vapor unless steps are taken to suppress the phenomenon or correct for it by demisting. Special surfaces that promote very rapid boiling are apt to be associated with a high level of entrainment.

On the other hand, ordinary smooth, metallic surfaces barely permit nucleation to occur under the small temperature potentials available in the solar sea power plant. Consequently the contribution of microconvection to the over-all heat flux across such smooth surfaces is generally negligible.

A viable means of producing a high heat transfer coefficient on the working fluid side of the heat exchanger area is to cause the evaporating liquid or condensate, as the case may be, to flow in a thin layer down a vertical surface under the section of gravity.

When such flow is applied to the solar sea power plant, almost no nucleation can be anticipated in the so-called boiler. Therefore, the boiler is really an evaporator and the processes occurring there are essentially the reverse of the processes occurring in the condenser. In either unit, the heat transfer coefficient is substantially influenced by the loading of the surface.

Flow Over Fluted Surfaces: A simple way to enhance condensation heat transfer in layer flow over vertical surfaces is accomplished by cutting fine axial flutes in the wall. Surface tension pulls the condensate on the ridges into the valleys, thereby leaving a very thin film on the ridges. It has been observed that the heat transfer coefficient is often larger than predicted theoretically. This increase is correlated to the wave motion in the falling film. This enhancement of heat transfer by falling waves was found both in condensation and in evaporation.

The passing wave wets all parts of the flutes but after the wave goes by, surface tension pulls the liquid away from the crests of the flutes and toward the liquid still flowing in the rills. Thus, between waves there is only a thin film of liquid covering the crests of the flutes.

Since the resistance to heat flow in such a thin film is extremely small, the average heat transfer coefficient over the whole surface is notably increased. Part of the increase is attributable to the hydrodynamic action just described and part is due to the fact that fluting increases the active heat transfer surface per unit width of the wall. The efficiency of flutes in increasing the heat transfer coefficient of a falling liquid layer should depend on the characteristics of the waves. The nature of the waves depends on the properties of the flowing liquid and the loading of the surface. Since the waves are strongly influenced by the gross type of flow and since the various regimes of flow are characterized largely through the Reynolds number, it can be anticipated that the efficiency of axial flutes should be highly dependent on the Reynolds number.

In view of the fact that transitional flow is associated with large, frequent waves containing vortices, it is to be expected that peak efficiency of the flutes should be attained somewhere within the transitional range of Reynolds numbers. For water-like fluids, at least, this is apt to be at Reynolds numbers between 2,000 and 4,000.

Extension to Other Liquids: In the absence of badly needed experimental data, the condition of peak efficiency on axially fluted surfaces carrying liquids other than water can be specified only tentatively. By way of a first pass at the matter, the peak might be expected to lie in the transitional range, not far from a Reynolds number of 3,000. The percent enhancement by virtue of increased area on the one hand and hydrodynamic wave action on the other can only be taken equal to the case of water until further data are available.

On the same tentative basis, the flute height can be related to the smooth surface liquid layer depth. The smooth surface depth is calculated for the liquid of interest at 3,000 Reynolds number. The cross section available for liquid flow is obtained from the geometry of the proposed flutes. When the calculation is performed for flutes of constant radius the necessary height of the flutes is 0.019 inch for ammonia at 70°F and 0.022 inch for propane at the same temperature.

UNIVERSITY OF MASSACHUSETTS DESIGN

The information in this chapter is excerpted from the following publications:

PB 228 067
PB 228 070
PB 236 422
PB 238 572
PB 239 369
PB 239 371
PB 239 373

OVERALL DESIGN CONCEPT (MARK I)

The UMASS SSPP is designed to be placed in the Gulf Stream 15 miles (25 km) east of the Collier Building of the University of Miami (Site One). A schematic of the design is shown in Figure 4.1. The temperature profile at Site One along an eastward extension of that line has been defined and tabulated below.

	For Site One, about 25 km east of benchmark				For Site N, about 31 to 50 km east of benchmark			
	Apr.	Aug.	Oct.	Dec.	Apr.	Aug.	Oct.	Dec.
1. Depth to Bottom	357m				686 to 802 m			
2. Selected Depth for Cold Water Inlet	330m				600 m			
3. T,Hot, in	77.0F	78.4F	80.6F	79.0F	77.6F	80.3F	81.9F	77.4F
4. T,Cold, in	44.9F	43.2F	49.2F	43.2F	43.4F	43.2F	44.8F	42.5F
5. ΔT	32.1F	35.2F	31.4F	35.8F	34.2F	37.0F	37.0F	35.0F

Source: PB 228 067

Design Conditions: T,Cold,in = 49F
ΔT = 31F

FIGURE 4.1: PROPOSED 400 MEGAWATT SEA THERMAL POWER PLANT

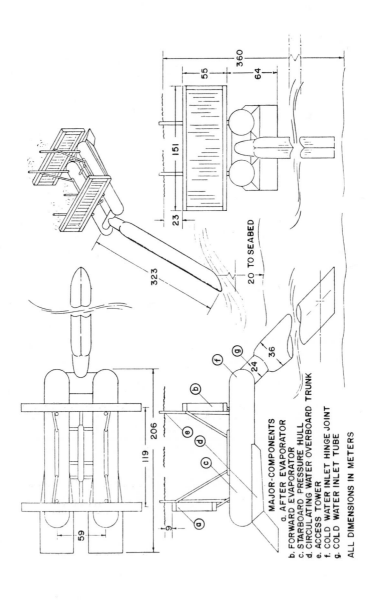

MAJOR-COMPONENTS
a. AFTER EVAPORATOR
b. FORWARD EVAPORATOR
c. STARBOARD PRESSURE HULL
d. CIRCULATING WATER OVERBOARD TRUNK
e. ACCESS TOWER
f. COLD WATER INLET HINGE JOINT
g. COLD WATER INLET TUBE

ALL DIMENSIONS IN METERS

Source: PB 228 067

These data indicate real advantage in moving out to eastward into deeper water. It remains to trade off those higher ΔT values against additional cold water suction pipe length and energy umbilical length. The system assumes a thermal working gradient of 31.5°F and an efficiency of 1.8%.

The Mark I is a large powerplant, 400 Mw_e net output. It is configured to ride at anchor in the flowing Gulf Stream: the ability of the Gulf Stream to replenish the hot water supply and to assist in replenishment of the cold water supply is an inherent part of this Ocean Thermal Gradient Electricity Generating System (OTGEGS). It therefore follows that the ability to maintain position and to transmit energy from the afloat plant to the shore are also inherent parts of this OTGEGS. The study therefore indicates use of an in-sea-bed energy collection grid, cable and/or pipe line, with connection nodes at anchor points, and transmission cables or pipes of at least 15 miles length extending back to the shore.

Each of the 400 Mw_e powerplants is configured as a semisubmersible ship, capable of operating at reasonable draft on the surface but usually lying in static balance at the end of its mooring with its mass so far beneath the surface that even hurricane season wind-wave motion excitation will be very small. The mooring line incorporates the energy umbilical and the cold water inlet pipe is a part of the mooring line. Present hull configuration is that of a pair of 100 foot diameter cylinders, side-by-side, each 600 feet long.

The evaporators are carried on top of the hulls to minimize ducting required to bring hot water to them, because they must receive water from the upper 20 meters of depth if possible, and they must not receive oil film from those near surface waters. The turbines and condensers are carried inside the pressure hulls. There is a multiplicity of turbines because the ideal size and capacity turbine has to be used so that the required 90% or more turbine efficiency will be realized. The Mark I thus contains sixteen identical power modules, each delivering some 37.5 Mw of electricity of which 25 Mw are available to the energy collection and transmission system.

The design of the evaporators and condensers is based on a complete two-phase heat transfer theory which has been synthesized for transfer of heat into ammonia or propane in the 40° to 100°F temperature band. The theory now covers plain tubes, tubes with extended internal or external surface and the very compact channels of the plate-fin geometry. The model includes all of the cycle pressure drops and parasitic pumping power requirements including those for the cold water supply. Overall system efficiencies between 1.5 and 1.8% using either ammonia or propane for the working fluid are predicted.

In all studies ammonia shows advantage over propane as the working fluid, but experience with ammonia refrigeration systems has shown the practical problems associated with ammonia to be so large that the less attractive propane will probably receive greater attention.

Analysis of the ammonia and propane turbines gives assurance of the required high efficiency. Some penalty is paid for allowing the turbines to take their ideal size, but the penalty is very small and certainly acceptable. One major aspect of the determination of the ideal turbine/power-module size is that it opens up ready access to a large family of powerplants whose net output can be some multiple of 25 Mw$_e$. For the Gulf Stream OTGEGS a total of sixteen of those power modules in one anchored envelope seems to be as large as one would care to cope with: but in the free-running high seas version of this concept a powerplant of 800 or 1200 Mw$_e$ is quite feasible and perhaps desirable. A plant using either ammonia or propane with a turbine smaller than 20 to 40 Mw shaft output will probably suffer drastically from inefficiency.

The resultant configuration, the Mark I at Site One, carries the evaporators on top of the hulls, with their liquid level considerably above the level of the condensate under the condensers (thermodynamically there is advantage to reversing that order for propane, but from a practical arrangement viewpoint it is not clear how that advantage can be realized). The turbines and condensers are carried inside pressure-proof one atmosphere hulls. The condenser concept makes possible the isolation of individual condensers for maintenance on either side and repair against leakage either by selective channel plugging or by core replacement from an on board stock of spares.

All of the condenser cores are pressure-proof. The cold water inlet pipe shown for Mark I is a very large open ended welded aluminum hull whose framing includes enough pressure-proof void spaces that neutral buoyancy can be achieved if desired. This pipe will permit as many as 30 million gallons of cold water per minute to traverse its 1,100 foot length while suffering very small function losses and minor hydraulic losses. Because of that short travel time there seems to be no need for thermal insulation to prevent warming of the cold water. A multiplicity of circulating water pumps, one or more for each condenser, is planned, high-volume low-head axial flow propeller pumps, another relatively modern type of rotating hydraulic machinery capable of achieving very high efficiency if properly designed.

It has been concluded that the biofouling which must be expected in these waters can be controlled, either by use of a high copper alloy metal in the heat exchangers, or by some combination of electrolytic chlorination, mechanical brushing or ultrasonic cleaning.

Cold Water Supply

The cold water available at selected Site One is not as cold as desired: this process would very much prefer 4°C water available elsewhere in abundance. The cold water in the proposed operating area seems to arrive there by movement from south to north under the surface current, by movement from east to west, and in the northern end of the swath, by countercurrent movement from north to south.

As to the structural problems and the hydraulics of the supply of cold water, it has been concluded that cold water inlet pipes of very nearly zero frictional resistance must be provided for this system to be economic, and that the form drag or miscellaneous losses in that hydraulic circuit must also be very small indeed. The solution to the problem has been identified as a very large cold water inlet pipe and a hydraulic path which can be approximated by one entrance loss, straight-line flow, one 90° bend, flow between condenser flat plates at velocities of the order of 2.25 ft/sec, and one exit loss.

Cold Water Suction Pipe

From a hydraulics point of view the length of suction pipe is of little significance. From the viewpoint of structural feasibility and cost of construction,
however, the length of that pipe is a major problem. Two different design
approaches have been taken to date in conceptualizing the pipe.

The first approach is one in which a pipe of circular cross section is to be assembled on site from prefabbed components, mostly tubular items, and is to
be lowered into the sea as its length grows until finally it is suspended in its
operating position. By controlling the buoyancy of the structure along the
length of the pipe, much of the dead weight of the structure is relieved, and
the pipe assumes some pendulous attitude of stability beneath the powerplant.
Suction pipe diameters of the order of sixty feet are under consideration.

The other approach has been more that of the design and fabrication of a
ship hull girder that would be completely fabricated in the building yard and
assembled in the ocean thermal powerplant prior to the transit from yard to
operating site. During transit and emplacement at the construction site the
tube would never see a significant torsional load. The structure is to have a
streamlined cross section of a foil shape known to be able to prevent the
shedding of Karman vortex sheets as the Gulf Stream current flows around it.
The first structure studied has a length of 1,100 feet, an internal diameter
across the flow of 80 feet (outside dimension), and a foil thickness ratio parallel to the external water flow of 0.445.

The loads analysis for this concept has treated the structure when on site in
use as an essentially axially loaded free-form member upon which a bending
moment caused by drag in the current has been superimposed. The current
pattern used is that for Site One with a factor of safety of 1.5.

The structure loading, when it is rigged horizontally as a nearly-awash hull,
has also been analyzed, crudely, to a first approximation, as a hull subjected
in either or both vertical and horizontal planes to the hogging and sagging
which the expected waves, Newport News to Miami, anytime of the year,
might create. These loads analyses were adequate to start the design but
require considerable refinement.

To overcome those loads a longitudinally and transversely framed welded shell
has been designed. The first design calls for 5,083 aluminum extrusions, plate
and shapes in the O condition. The depth of framing is such that the least
internal dimension across flow is now about 76' instead of 80'.

The longitudinal framing uses extruded cylinders for standing flanges sized and
built so that they will contribute enough pressure-proof displacement to the
structure to render it neutrally buoyant when awash. The shell and longitudinal framing are to be arranged so that the 16×10^6 lb of mooring tension
can be handled both generally and at the attachment details where the tether
will attach to the cold water suction tube and where the cold water suction
tube will attach to the powerplant hulls.

Turbine and Turbine System

The ammonia turbine is smaller and much less expensive than the propane turbine, but the propane turbine is still quite reasonable in cost. The probability
is that turbines with efficiencies in excess of 90% can be had for this application.

Naval Architecture

The naval architecture of the system, displacement and weight, resistance,
static balance, and the desirability of configuring the mooring tether to minimize the vertical component of force placed on the anchor has been studied.

The Mark I must have rather extensive capabilities as a submarine as well as
a surface ship. Those submarine capabilities can be described as the possession
of submerged neutral buoyancy, an adequate Main Ballast System that will
permit change from the surfaced condition to the neutrally buoyant submerged
condition, and an adequate Variable Ballast System that will permit accommoda-
tion in any of the force vectors which enter the free-body diagram.

Because there must be a considerable upward force vector in the free-body
diagram to overcome the downward component of force in the mooring, this
Mark I hull will not actually achieve the same kind of condition of neutral
buoyancy, submerged, that would be found in a free-running submarine. It
has therefore been necessary to define several conditions of Loading and Attitude, in which the Mark I will find herself at any time in the sequence from
building site to completion of emplacement on operating site, and to show
how ballast will be changed to go from one condition to another.

Several emergency situations with which the Mark I must cope have also been
postulated, and the requirements associated therewith set down in terms of
subsystem capabilities. The hull envelope shown in the Mark I mockup does
not yet include the volume required for Main Ballast Tanks. The resistance
force vector in the free-body diagram has been calculated assuming a worst-
case situation with the two evaporator faces taking the form of two solid flat
plates normal to the oncoming stream.

In addition to an assessment of those fundamentals of statics the dynamics of
the situation must be studied. It is essential that wind-wave excited motion be
understood and that any that will exist be acceptable. Concentration is being
placed on the heave motion because that will probably be the most trouble-
some: pitch and roll and yaw, first separately, then coupled will also be studied. There is also the requirement that the cold water pipe and any other
parts of the system be free from undesirable vibration: the cold water pipe
geometry must be selected most carefully to avoid undesirable vortex shedding
phenomena.

It is not yet clear what the actual geometry of the evaporators will be: they
will probably be located topside and will probably ride with their tops only a
few feet beneath the surface. The influence of wind-waves on them may become very significant, and it may be essential to provide for fine-tuning of
clearance over the tops of evaporators as sea state varies.

Partially uncovered evaporators would get into structural difficulty early on in rising seas and should therefore probably be adequately immersed. On the other hand, the system wants to be able to take advantage of the one or two degrees of increased temperature in the near-surface layers. Attempts have been made to sketch configurations that could be constructed in building basins or on shipways and still place the evaporators below the condensers with provision made for delivering the hot surface water to the evaporators.

Smaller systems of the order of 25 Mw_e (one propane power module) fit that configuration very well; but larger units, multiples of the most efficient propane power module, do not adjust nicely to that configuration. Topside evaporators appear at this point to be the most likely situation, and their effect on the system architecture is being given due attention.

The first analysis of Drag of the Mark I at Site One yielded a 20×10^6 lb force answer. A factor of 1.5 was applied to the velocity envelope at Site One, and worst-case assumptions were made for the evaporators and the possibility that extensive downward directing ducting will be wanted on the circulating water overboard discharge ports.

It is thought that the next configuration, the Mark II, will show a reduction from that 20 million pounds number. If the concept of using the moving Gulf Stream as supplier of hot and cold water is retained, the magnitude of the power that would be required to hold a powerplant on station by dynamic positioning as compared against anchored positioning can be seen as very large.

In fact, dynamic positioning out where the best ΔT exists in the Gulf Stream, would not be practical. In the deep water of the Gulf of Mexico, a minimum of 200 miles south of the Gulf coast, on the other hand, dynamic positioning would require practically no power at all. (For that site, mobile cryogenic hydrogen tankers and hydrogen liquefaction equipment necessary to bring the energy to market must be investigated because neither cable nor pipeline would appear to work there.)

Anchor and Mooring System

Studies of the seabed soils, their holding capabilities, and candidate anchor and mooring systems for the Mark I at Site One and at thousands of similar southeast coast Gulf Stream sites, show the concept of mooring this system to a single anchor to be feasible but formidable. There are the broadest variety of seabeds within the proposed deployment area, from bare rock to unconsolidated sand. One type of anchor, the simple gravity anchor, will suffice for any of those conditions, and a gravity type anchor of reasonable cost and requiring construction methods no different from those already in use has been conceptualized.

The anchor is to be made from reinforced concrete and will take the form of a concrete barge hull about 300' long by 60' broad by 30' deep. The interior compartmentation of the anchor will be arranged so that it will float at about 15' draft when transiting to the emplacement site; there it will become a submarine, capable of diving in controlled manner to the seabed where it will rest exerting some 34 million pounds of reaction downward. There will be pressure-proof compartments inside the massive concrete hull which will serve as connection points for both electricity and gas pipeline energy collection sub-systems, and into which the energy umbilical will be connected.

Emplacement of these anchors with precision on a predetermined grid will be started using an auxiliary emplacement anchor and distance line, but subsequent downstream anchors can be emplaced using the anchor immediately upstream in the grid as the emplacement controlling device. Cost of high quality concrete anchors in place has been estimated to be of the order of $30/kw, assuming a rating of 400 Mw$_e$ and taking into account the significant effort that will be required for this entire evolution. The gravity type anchor is but one of many possibilities: piling driven into templates might yield anchor subsystems of cost lower than that of gravity type.

A single mooring line from anchor to powerplant has been selected as the most desirable configuration. The Mark I system actually uses the axial strength of the cold water inlet pipe (CWIP) as a major portion of its mooring line.

The tether between the CWIP and the anchor has been given the configuration of a submarine in itself: the proposed configuration achieves the strength and simplicity of design required, permits the tether to be built and moved about as a floating system, and allows it to be emplaced in a controlled diving mode which will carry it to the seabed and leave it there in a condition which reduces the vertical force component on the anchor if so desired. Fabrication of the tether from Monel clad high strength steel forgings and wrought pieces has been proposed, with a total electrically insulated joint provided between it and the CWIP if made from welded aluminum.

The anchor and mooring system conceptualized matches very well the concept of grand scale practice of the ocean thermal differences process along the southeast coast and the concept for an energy collection grid, cable and pipe-line, placed in the seabed, feeding into the U.S. electricity and/or hydrogen gas fuel market.

Energy Umbilical

This project has identified one system for detailing, the Grand-Scale System comprising large numbers of powerplants, each as large as 400 Mw$_e$, all afloat in the moving waters of the Gulf Stream, moored from the anchor point nodes in an extensive energy collection system. The energy umbilical between the powerplant and the energy collection system takes the form of submarine elec-tric cables fastened inside the cold water inlet pipe, and/or 300 psia working pressure hydrogen pipeline and hose, similarly fastened within the pipe.

At the powerplant end of the umbilical both electricity and gas must leave the pressure-proof hull and pass into the CWIP via swivel joints. At the bottom of the CWIP both electricity and gas must be connected via waterproof con-nectors to the tether: the portion of the umbilical that passes through the center of the tether must possess adequate flexibility. The umbilical at the anchor end of the tether must enter the splicing hull built into the anchor, the compartment within which connections are made to the seabed energy collection system.

The DC electricity umbilical with the basic requirement that at least 400 Mw be transmitted for a distance of at least 30 miles has been studied. The subsystem starts with AC generators, AC to DC rectification using mercury vapor valves, then to submarine DC cable. Recent seabed pipeline construction technology indicates the feasibility of laying of seabed pipelines in 1,600 to 3,000 ft of water beneath a considerable current.

Control of Biofouling

If system economics will permit the use of 90/10 copper-nickel alloy in the heat exchangers, the inherent antifouling characteristic of that material plus the 2.25 ft minimum per second seawater flow velocities required in evaporator and condenser cores will eliminate concern about biofouling. If aluminum or titanium or carbon loaded plastics are used for exchanger surfaces, one or more of the following techniques in combination will be used to control biofouling:

(a)   chronic chlorination of exchanger surfaces, preferably accomplished by electrolytic release on surfaces themselves,

(b)   periodic chlorination by displacement of seawater within isolated exchanger boundaries by chlorinated water prepared and stored in batch tanks for that purpose,

(c)   mechanical brushing using rotating brushes in the water passages, brushes made to move up and down to cover the total plate surface, or

(d)   continual or periodic application of ultrasonic vibratory forces to the heat exchanger plates.

The efficacy of each of these methods should be demonstrated full size in the same test facility used to verify the heat exchanger analysis and design theory. The use of antifouling coatings on heat exchanger surfaces has been suggested, but it is feared that any such coating, no matter how carefully and thinly applied, would reduce metallic transfer coefficients considerably. It has also been suggested that copper wire spirals or screen in the outer surface of plastic exchanger surfaces could provide the same protection which one has with the high copper alloy surfaces.

HEAT EXCHANGER DESIGN (MARK I)

The magnitude of the heat exchange task can be approximated from fundamental considerations. Assuming a 100,000 kw plant output and 2.5% thermal efficiency, then

$$\text{Heat input} = \frac{100}{2.5} \times 100,000 \times 3,413 = 1.365 \times 10^{10} \text{ Btu/hr}$$

$$\text{Heat rejected} = \frac{100 - 2.5}{2.5} \times 100,000 \times 3,413 = 1.331 \times 10^{10} \text{ Btu/hr}$$

The heat flow equation is as follows.

$$Q = UA\Delta T$$

Where Q = Btu/hr heat transferred

U = Overall heat transfer coefficient $\sim$Btu/hr ft² °F

A = Area of exchange wall in ft²
(This assumes both sides to be the same area)

ΔT = Temperature difference between hot fluid and cold fluid $\sim$°F. Where temperature of the fluid changes as it passes through the exchanger. ΔT is usually assumed to be the logarithmic mean temperature difference, or LMTD.

The assumption where U = 200 Btu/hr ft² °F for both boiler and condenser, should be a reasonably achievable value. Assuming warm water enters at 82°F and leaves at 79°F, and fluid boils at 72°F,

$$\Delta T = \text{LMTD} = \frac{(82 - 72) - (79 - 72)}{\ln[(82 - 72)/(79 - 72)]} = 8.411°\text{F}$$

From these values, and the heat flow the boiler area can be computed.

$$A = \frac{Q}{U\Delta T} = \frac{1.365 \times 10^{10}}{200 \times 8.411} = 8.11 \times 10^6 \text{ ft}^2$$

Assuming cold water enters the condenser at 43°F and leaves at 49°F and fluid condenses at 52°F, then

$$\text{LMTD} = \frac{(52 - 43) - (52 - 49)}{\ln[(9)/(3)]} = 5.461°\text{F}$$

$$A = \frac{1.331 \times 10^{10}}{200 \times 5.461} = 12.19 \times 10^6 \text{ ft}^2$$

Adding boiler surface area to condenser surface area, a total area equal to $(8.11 + 12.19) \times 10^6$, i.e., 20.30×10^6 ft² is obtained.

Large conventional tubular heat exchangers are likely to cost $4.00 to $10.00 per square foot (1973 $s). At this price a first approximation of heat exchanger costs for a 100 Mw plant would be $81 to $203 million, or $812 to $2,030 per kilowatt.

Obviously the heat transfer cost is probably the biggest single cost in a Sea Thermal Plant, and it therefore is essential to explore every possibility for reducing this cost.

Exchanger Requirements

Heat Flow: As pointed out above heat flow is very large, and is directly affected by efficiency. For example, if efficiency is improved by 5 to 2.5% the total heat flow is affected as shown on the following page.

$$Q = Q_{in} + Q_{out} \text{ (power out)}$$

$$= (1/e + 1/e - 1) \text{ or } 2/e - 1$$

$$= 2/0.025 - 1 = 79.0$$

and $[2/(0.025 \times 1.05)] - 1 = 75.19$ giving a ratio of 79.0/75.19 or 1.0507.

Improving turbine or generator efficiency by 5% reduces heat flow and heat transfer by an even greater ratio making it very important to do almost everything possible to improve plant efficiency, because it will reduce heat flow and heat transfer cost.

Temperature Difference: It is obvious that increasing the temperature difference across the heat exchangers would decrease the area requirement. However, within the available temperature limits this reduces the temperature difference between boiling and condensing temperature, which in turn reduces plant efficiency and increases heat flow required. For this reason there must be an optimum economic temperature difference that should be used for best plant economics.

One direct means for increasing temperature difference is to extend the cold water pipe to deeper and colder waters. However, the rate of change of temperature with depth decreases rapidly below approximately 800 to 1,000 meters, as shown on Figure 4.2. The deep water pipe costs more as it goes to greater depth, and therefore there is some optimum economic limit to the pipe depth. With economical, strong, and flexible pipe construction, it is quite probable that the pipe depth can be extended to reach water temperatures of about 40° to 39.2°F, instead of 43°F as first assumed.

Multiple Circuits: One important way to increase the effective temperature difference, or LMTD is to use multiple fluid circuits in parallel. This can be illustrated by taking the same water conditions assumed above and calculating LMTD for two fluid circuits. The mean boiling temperature is 72°F, but by using two fluid circuits in the condenser two boiler temperatures of 72.75°F and 71.25°F are obtained. Under these conditions

$$LMTD = \frac{(82 - 72.75) - (80.5 - 72.75)}{\ln[(9.25)/(7.75)]} = 8.478$$

The ratio of this to the LMTD in the single circuit case is 8.478/8.411 or 1.0080.

In the condenser two condensing temperatures of 53.5° and 50.5°F instead of the mean temperature of 52°F can be used, then

$$LMTD = \frac{(50.5 - 43) - (50.5 - 46)}{\ln[(7.5)/(4.5)]} = 5.873$$

Ratio to that for the single circuit is 5.783/5.461 or 1.0754.

FIGURE 4.2: RATE OF CHANGE OF TEMPERATURE

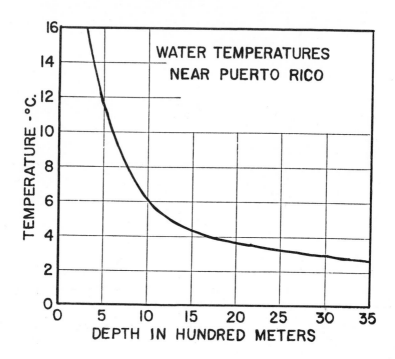

Source: PB 239 369

Calculation of ideal cycle efficiency indicated a very slight improvement in cycle efficiency attained by the use of multiple fluid circuits. Because of the efficiency improvement there is also a very slight reduction in heat rejection load.

While these latter two effects are negligible, they illustrate that there are three benefits from using multiple fluid circuits:

   (1)   It increases mean temperature difference acorss the heat ex-changer,

   (2)   It improves cycle efficiency, and

   (3)   It reduces heat rejection load.

The temperature drop in the warm water through the boiler is preferably less than the temperature rise in the cold water through the condenser, the reason being that it is much more costly to bring cold water than to bring warm water to the plant. The effect of higher temperature rise of the condenser water is to reduce available LMTD, but this effect is alleviated by using multiple fluid circuits.

Since the optimum temperature drop on the boiler water will be lower than that on the condenser there is not as much advantage to making multiple fluid circuits on the boiler as on the condenser. Therefore a desirable compromise may be to use 2 circuits on the boilers and 4 on the condensers. This arrangement would fit in well with double flow turbines, each having single inlets and two separate exhaust lines.

Boiling Heat Transfer: The boiling of the fluid in a Sea Thermal plant is most nearly like that in an evaporator for a water desalting plant. Temperature differences are much lower than those in an ordinary steam boiler, and boiling intensity is much lower. The temperature differences are also very close to those in a typical refrigerant evaporator.

There is one distinction between evaporation in the usual refrigerant evaporators and in the Sea Thermal fluid boiler. In the usual refrigerant evaporator the fluid has a high pressure drop through an expansion valve. The temperature of the liquid entering the expansion valve is higher than the boiling temperature.

Therefore the liquid flashes into a high volume percentage of vapor as it leaves the expansion valve. It also increases local velocities very greatly and increases the boiling heat transfer coefficient. Contrasted to this in the Sea Thermal boiler the liquid enters at approximately 20°F below boiling temperature and it must be heated before it can boil. This makes the heat exchange problem more difficult than that in a refrigeration cycle evaporator. A flooded evaporation with liquid recirculation evaporator is most probably the best type system for a Sea Thermal boiler.

Condensing Heat Transfer: Heat transfer from a condensing vapor to a cold surface is generally excellent, and can be one of three types: it can be dropwise, through a laminar liquid film, or through a turbulent liquid film. The most common type, and that likely to be encountered in Sea Thermal condensers is laminar film type.

Water Side Heat Transfer: Heat transfer between water and a surface is well established and understood. Normally, high surface heat transfer coefficients are easily attainable with water by increasing water velocity. However, a Sea Thermal plant is somewhat different, because so little heat is taken out of the warm water, or added to the cold water. Therefore, large quantities of water must be pumped through the exchanger.

Since the friction head loss is approximately proportional to velocity squared and since pumping power is a direct loss in the plant, it is important to keep the velocity rather low. Preliminary data has shown that velocities should be kept down to approximately 2 ft/sec for best overall plant economics. This velocity is much below the usual values of 6 to 10 ft/sec used in tubular heat exchangers, and the heat transfer coefficients are correspondingly lower.

The second factor vitally affecting water side heat transfer is the film caused by dirt, organism growth, scaling or corrosion buildup. A Sea Thermal plant will definitely have this fouling problem to cope with on the water side. If halocarbons or hydrocarbons are selected as fluids, there should be no fouling film on the working fluid side.

From these general observations, and evaluation of the resistance in the water film and fouling film, it soon becomes apparent that the water side heat transfer resistance is the major one and largely controls the heat exchanger design. While it is obviously desirable to develop high boiling and condensing coefficients, both are likely to be small resistances compared to the water side resistance.

Marine Fouling: Coating of heat transfer surfaces with various marine organisms is a possibly serious problem in Sea Thermal heat exchangers. Traditionally copper bearing alloys are used to prevent this by being poisonous to these organisms. It is also well-known that maintaining high water velocities tends to inhibit growth and attachment of these organisms to surfaces.

The use of copper alloys may be too costly for use in a Sea Thermal plant. The use of high water velocities may also be too costly because of high friction losses. Whether the velocities that are economically feasible in a Sea Thermal plant are sufficient to prevent fouling is open to question. Therefore, fouling might be expected to be a serious problem. Extensive experience with seawater usage suggests continuous chlorination to produce a residual value of 0.5 ppm of chlorine. It has also been suggested that as little as 0.25 ppm are effective.

It has been estimated that this would cost 2.92¢/kwh (1973 $s), which is prohibitively high, and obviously not a practical solution to the problem. It has been suggested that hypochlorous acid could be generated by direct electrolysis of seawater. This should reduce the cost somewhat, but it is doubtful if this cost could be reasonably low.

The most logical solution to the fouling problem is probably periodic cleaning of the water side heat transfer surface. This can be done chemically by closing the water side and soaking it in higher concentrations of chlorine. This means that the heat exchangers must be designed for easy cleaning, either chemically, hydraulically, or mechanically. It also probably means that complicated, or highly specialized designs of heat transfer surface for improving water side heat transfer are practically useless.

Corrosion: Corrosion is a traditional problem in seawater heat exchangers and must be expected to be a serious one in Sea Thermal plants. There are several alleviating factors, which may make this problem a little less severe in a Sea Thermal plant.

(a)   In the usual case seawater is used as a coolant for power plant condensers or process coolers, and is heated 10°F or more above natural temperature. Corrosion is increased at the higher temperatures. In the Sea Thermal plant the warm water temperature drops as it passes through the boilers. This should decrease corrosion, compared to common usage.

(b)   The cold water to the condenser is not likely to be as corrosive as the warm surface water, simply because corrosion is lessened at lower temperatures.

(c)   Water in the open sea is generally cleaner and less corrosive than harbor water, or most coastal waters.

(d)   Conditions of temperature, flow, and water are likely to be more

nearly constant in a Sea Thermal plant than they are in a ship
traveling through many waters, or in a coastal industrial installation.

Many thousands of ships and industrial installations for using seawater are in
successful operation to prove that corrosion is being coped with at a price.
Most of these installations are operated under more difficult corrosive conditions than those in which Sea Thermal heat exchangers will operate. It therefore can be said that the corrosion problem can be handled.

The real challenge lies in the cost of the solution. Since the heat exchangers
are so very large it is questionable if expensive materials such as titanium,
Monel, cupro-nickel, or high alloy stainless steels can be afforded. Probably
the only possibility of using these costly materials would depend on using very
thin walls, so that not so much material is needed as in a conventional heat
exchanger.

In the development work on water desalting much progress has been made toward using aluminum and plastics for heat exchange surfaces, and there is good
reason to believe that one of these low cost materials for exchanger surfaces
can be used.

Debris: In a Sea Thermal plant the requirement for a large amount of surface
automatically necessitates making small water passages, so as to pack the transfer area into the smallest possible volume. Small passages are much more sensitive to plugging with small objects, such as fish, weeds, shrimp, etc.

Therefore, it becomes important to develop a good screening system to protect the exchangers. While this is common to most cooling systems, the difference here is that at least the cold water screens must be submerged in the
ocean, and must be easily cleanable and maintainable. There is no reason to
believe this cannot be done, but it must be carefully considered in the design
of a plant.

Leaks: It would be naive to think that in a Sea Thermal plant, with such a
large heat transfer area, and with many interconnecting pipes required, the system could be made leakproof. The very fact that low costs must be achieved
in order to make the plant economically feasible also tends to make leaks more
probable. To compound the problem, it is self-evident that leaks will be more
difficult to repair in an undersea environment than in the open atmosphere.

The inevitable conclusion is that leaks in a Sea Thermal plant are unavoidable.
It is probably advisable to keep the pressures in the heat exchangers slightly
lower than the surrounding water pressures so that water will leak into the exchangers, rather than have system fluid leak out into the ocean. Water leak-
ing into the exchangers can be separated from the system fluid, provided that
the system fluid is not soluble in the water.

It should also be noted that by keeping water pressure as close as possible to
system fluid pressure leakage is minimized. This is an important reason for
submerging heat exchangers to a depth where water pressure is equal to system
pressure.

Materials

Various materials are possible for use in construction of the heat exchangers. The desired properties are good corrosion resistance, good conductivity, formability, sufficient strength, and low cost. Since it is doubtful if any one material will have all of these properties, a compromise selection must be made.

Table 4.1 shows some values for the properties of various possible materials. All of these materials are potentially suitable for corrosion resistance. If tubular construction is used then it is doubtful if walls of tubes can be manufactured and assembled in thicknesses less than 0.030" thick. If flat sheet construction is developed, and pressures are equalized between working fluid and water, then it is quite conceivable that wall thicknesses as low as 0.006" can be used.

TABLE 4.1

Material	K, Btu $\frac{ft}{hr}$ ft^2 F	Tensile Strength, psi	Cost/lb, $	Density, g/cc	Cost/in^3	Cost/ft^2, ($), 0.006" Thick	Cost/ft^2, ($), 0.030" Thick
Titanium	9.5	35,000–85,000	7.55	4.37	1.19	1.03	5.14
304 Stainless	9.4	85,000	0.93	7.83	0.263	0.227	1.14
70–30 Cu-Ni	18	70,000	–	8.93	–	–	–
90–10 Cu-Ni	22	60,000	–	8.93	–	–	–
Aluminum Alclad	102	22,000–39,000	0.52	2.73	0.051	0.044	0.22
Polyvinyl fluoride	0.10	7,000–18,000	2.50	1.57	0.142	0.123	0.61
Polyethylene	0.19	2,000–4,000	0.20	0.93	0.0067	0.0058	0.029
Polypropylene	0.14	3,000–10,000	0.30	0.91	0.0099	0.0086	0.043
Carbon polyethylene	0.30	3,000–5,000	0.25	1.0	0.0090	0.0078	0.039

Source: PB 239 369

Pertinent properties shown are approximations, and properties of specific alloys or compositions will vary slightly from those shown. Costs also are representative approximations.

It should be noted that the costs are for the material in the exchange surface only. They do not include materials for the structure, fluid enclosure, and connections, nor do they include manufacture, fabrication, and assembly costs. Thus, Sea Thermal heat exchangers will need to be carefully designed for ease and simplicity of manufacture, as well as for heat exchanging effectiveness.

Getting the most heat transfer per dollar invested, which includes not only the heat exchanger but also the cost of pumping equipment and powerplant to overcome the friction loss in the heat exchangers is the primary goal in the design desired.

Analysis indicated that a heat exchanger section like that shown on Figure 4.3 would be a good cost effective design. This consists of sandwiches of two flat sheets bonded together by a corrugated separator with corrugations providing the vertical path for the condensing vapor and liquid in the downflow direction. For the boiler the flow of liquid and vapor would be upward through the corrugated passages. If the corrugations are made of aluminum, with high thermal conductivity, then these sheets provide extra heat transfer surface on the vapor side, just as fins do on a finned tube surface.

FIGURE 4.3: HEAT EXCHANGER

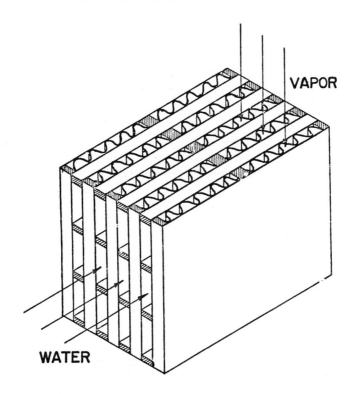

Source: PB 239 369

This extended surface is only effective if the flat sheet surface is also quite conductive, such as would be the case if both sheets and corrugations were aluminum. However aluminum corrugations would be effective in insuring full use of the flat sheet surface, even if low conductivity plastic flat sheets were used. The aluminum corrugated sheet can be perforated in various possible ways to promote turbulence in the boiling, or to shorten the liquid film path on the condensing side.

The water passages are long rectangles in cross section, are cleanable and simple, and can easily be designed for optimum flow and transfer conditions. As shown on the drawing, it is possible to divide the boiling or condensing fluid side into multiple channels to provide better heat transfer effectiveness.

The sandwich of two flat sheets with corrugated material between can be made by a corrugating machine similar to that already being used to manufacture paper boxboard. Corrugated box makers have already made board with paper sheets bonded to each side of a steel or aluminum corrugated sheet. The process for making corrugated paperboard uses three rolls of paper feeding through the machine at speeds up to 1,500 fpm. The middle sheet is corrugated by feeding it through two toothed rolls, and the two outer sheets are fed through the machine while being bonded continuously to the inner corrugated sheet.

It should be possible to make these machines to produce corrugated plastic sandwich sheets in the same way, with either plastic or metal corrugated sheets between the flat plastic sheets. Bonding can be done either by applied adhesive or by heat bonding.

Aluminum heat exchangers are already in mass production for various purposes, made according to these principles. In this case the aluminum sheets are stacked together, and then furnace or salt bath brazed to form a complete exchanger assembly. This construction permits a rather low cost design. The major problem with the all aluminum brazed construction is that it is difficult to braze very large assemblies, and sheet thickness less than 0.030" cannot be used. With adhesive bonding exchangers with sheet thicknesses as low as 0.006" should be possible. The advantage of this is that the wall resistance is reasonably low, even for poor conductors like plastics.

Detailed performance and cost analysis has shown that costs are reasonable enough so that they would present no barrier to successful economics for Sea Solar Power. Aluminum appears to be the most economical material for the exchangers.

Although aluminum appears to be the best material, should some unforeseen defect appear to prevent its use, then the more costly metals could be used and costs would still be permissible. Even the most costly titanium could be used, if absolutely necessary.

Plastics could provide low enough heat transfer sheet cost to be economical, but the necessary larger size of the exchangers and the high friction losses rule them out for the present. However, the development of plastics with higher conductivity may eventually permit their use.

The sheet thickness of 0.006" which was used in the cost analysis can probably only be assembled by adhesives. Brazing assemblies presently require a minimum sheet thickness of 0.030". While this would increase the cost of sheet and fin material, multiplying these costs by 5 would still allow costs low enough to be economically feasible.

Water side coefficients are generally the most important controlling factor in the overall heat transfer resistance. This is partly because of the inevitable fouling on the water side, and partly because it is very easy to put extended surface on the boiling or condensing side.

Surface heat transfer coefficients are presently high enough so that it is practical right now to build suitable exchangers for this service, without research and development toward better coefficients.

However, the immensity of heat exchanger requirements for this surface should not rule out the desirability of developing better coefficients with low attendant friction coefficients.

Optimum water velocities appear to be high enough to prevent marine fouling, however, it is probably advisable to set water velocities a little higher than the apparent optimum so as to minimize the fouling problem.

Compacting the surface so as to get minimum casing volume appears to be more important to the achievement of low cost than the development of higher transfer coefficients.

VARIATIONS IN HEAT EXCHANGER DESIGN (MARK II)

This study applies specifically to a closed cycle Rankine cycle engine with pressure-proof heat exchangers based in the Gulf Stream. Table 4.2 summarizes a preliminary estimate of specific capital costs of major system groups for this system, the Mark II design.

TABLE 4.2: COST ESTIMATE FOR 400 Mw POWERPLANT (MARK II)

Description	- - - Cost, $/kw - - -	
	Low	High
Hull and concrete components	13.5	39.0
Cold water intake pipe	45.0	63.0
Machinery	136.0	179.0
Heat exchangers	140.0	340.0
Electrical and transmission	44.0	82.0
Command and control	0.5	0.5
Auxiliary/life support	5.0	5.0
Outfit and furnishings	3.1	3.1
Total	387.1	711.6
Present value of construction with interest and escalation	433.6	797.0

Source: PB 238 572

Although the values reported are subject to major change, it can easily be seen that the heat exchangers are the most important cost component of the system. Since the estimated costs for heat exchangers are a strong function of material requirements, it is important to minimize their size via analytical design. Furthermore, since the cost of key components such as turbines, cold water delivery pipe, and hull containment system are influenced or dictated by heat exchanger design, it is doubly important to optimize the design of the boiler and condenser components.

The improved model for the heat exchangers along with the total cycle analysis program have formed the analytical basis for the latest powerplant configuration (identified as the Mark II system) which is shown in Figure 4.4. The hull and cold water delivery system are similar to the previous Mark I system, however, the use of plate-fin exchangers has greatly reduced the size of the boilers. Another major difference in the two systems is the change from naturally pumped boilers (using the kinetic energy of the Gulf Stream) to mechanically pumped heat exchangers.

FIGURE 4.4: MARK II SYSTEM

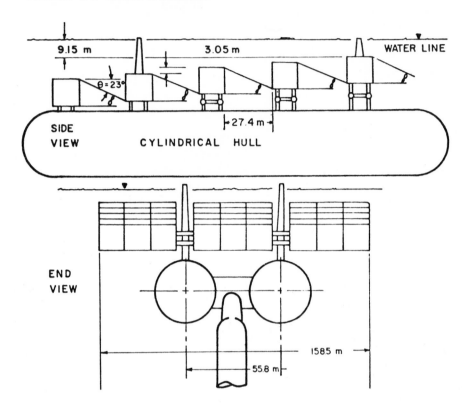

Source: PB 238 572

This change is based on the results of another study, that specifically studied the naturally pumped system configuration. The system as shown, is based on propane as the working fluid and consists of 15 top-side mechanically pumped evaporators arranged in 5 levels of 3 boiler modules each supplying 26.7 Mw of net power. For economic and damage control reasons, the boilers are arranged in common manifolds of 3 boilers each. Inside the hulls are placed 16 power modules (turbine, condenser, and working fluid pump) of 25 Mw each. The choice of power module size was dictated by system turbine requirements.

Condenser Placement

One method proposed to eliminate the problems of the cold water delivery and large ocean water pumping requirements is to place the condenser, or the entire power system, except boiler or turbine, far below the surface in the cold water sink. An investigation to determine the feasibility of this proposal was carried out for the system arranged as shown in Figure 4.5.

FIGURE 4.5: SYSTEM FEASIBILITY ARRANGEMENT

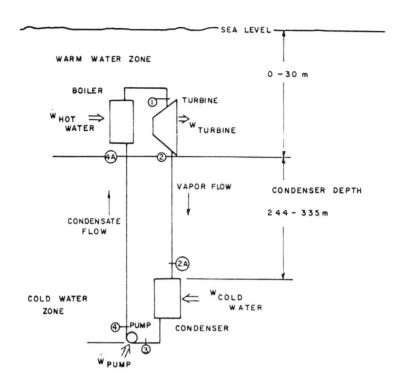

Source: PB 238 572

The total cycle program was modified to model this system, specifically to include the additional cycle pumping work (or gains in pressure head), and the heat exchangers were designed to withstand the increased hydrostatic pressures.

One comparison of this system (condenser submerged to 305 meters) with a conventional ammonia cycle arrangement is given in Table 4.3. This result clearly shows that the savings in cold water pumping are offset by increased cycle parasitic pumping demands. Although not as large as the required cold water delivery tubes, the vapor and liquid working fluid delivery tubes are quite large and need to be designed to withstand extreme hydrostatic pressures.

However, in spite of the additional strength requirements, the specific material requirements for the condenser core (based on the use of aluminum) in the submerged system are less than the conventional system, because of slightly higher overall heat transfer coefficients. For a fully equipped condenser which would include headers and other flow ducts capable of withstanding the extreme hydrostatic pressure, it is doubtful that this small specific weight advantage would stand.

TABLE 4.3: COMPARISON OF SUBMERGED CONDENSER AND CONVEN-
TIONAL SYSTEMS (400 Mw Net Power Output)

	SUBMERGED CONDENSER SYSTEM	CONVENTIONAL CYCLE ARRANGEMENT
Boiler Material ($\frac{Kg\ Aluminum}{Kw}$)	11.5	9.8
Condenser Material ($\frac{Kg\ Aluminum}{Kw}$)	12.3	12.7
Cold Water Tube Diameter (m)	-	24.4
Vapor Delivery Tube Diameter (m)	12.2	-
Liquid Delivery Tube Diameter (m)	7.6	-
Total Parasitic Power, Mw	143	119
Boiler Pumping	17.8%	17.4%
Condenser Pumping	25.2%	24.2%
Cold Delivery Tube Pumping	0	53.0%
Cycle Work	57.0%	5.4%

Source: PB 238 572

Because of the need for the two large pressure-proof working fluid delivery
tubes instead of one, much larger, and nonpressure-proof cold water delivery
tube, and the increased cycle pumping work, investigation of this type of cycle
arrangement was ceased. However, there are other potential cycle arrangements
that are worthy of investigation. One such arrangement for a pressure balanced
heat exchanger configuration, is to place the condenser above the boilers in
respect to the ocean surface. In addition to reduced exchanger material requirements, the cycle pump work may be reduced (or completely eliminated
for certain working fluids).

Boiler Ocean Side Pumping System

The first complete ocean power system designed from the early analytical
model, the Mark I system, was based on the use of the kinetic energy of the
Gulf Stream to pump hot working fluid through the boilers. In the original
staggered tube boiler design, a large fraction of the Gulf Stream flow (approximately 2 m/sec) was assumed to flow through the two large boilers placed
perpendicular to the Gulf Stream flow.

An analytical study characterizing the flow field around a Gulf Stream pumped
boiler and a combination with the total cycle model has enabled a more precise determination of the hot ocean water flow through such an arrangement.
This work has shown that the possibility of having a forward and aft evaporator
was not realistic because the forward evaporator removes most of the Gulf
Stream's kinetic energy.

However, for this study, the possibility of a natural pumped system with one large boiler (or group of boiler modules) has been investigated in detail. As an example of the analytical results, Figure 4.6 shows the amount of boiler core material for a 25 Mw ammonia system module as a function of evaporator core depth and specific working fluid flow rate.

FIGURE 4.6: AMMONIA BOILER 25 Mw NET POWER

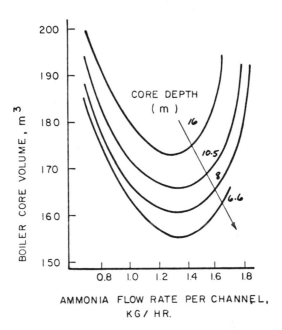

Source: PB 238 572

As a comparison, a similar ammonia system using a forced pumping system has a much smaller exchanger material volume requirement of approximately 100 cubic meters. Also, limiting the overall width of the single row of boiler modules to a minimum value of about 150 meters can place an upper bound on the maximum powerplant size as shown in Figure 4.7. From this graph it can be seen that, using the present hull size and a maximum boiler width of 150 meters, maximum net powerplant sizes of only 200 or 300 Mw are possible for propane or ammonia systems respectively.

It appears, that the savings in reduced hot water pumping work and capital costs are not enough to offset increased heat exchanger material costs and costs due to smaller total powerplant size.

Heat Exchanger or Power Module Size

Obvious factors as hull design or overall powerplant size restrictions can limit the maximum size of heat exchanger module size, however, restrictions on

FIGURE 4.7: MAXIMUM POWERPLANT SIZE

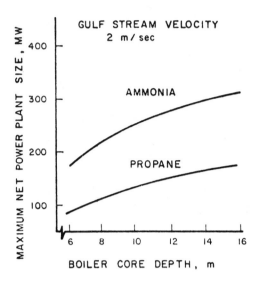

Source: PB 238 572

turbomachinery design can also directly influence this parameter. Based on the results of a detailed design and economic analysis of the turbomachinery components, an optimum size for each turbine power unit can be determined.

Results illustrating this point for single stage ammonia and propane turbines are shown in Figure 4.8. For propane, it can be seen that designing each turbine or power module close to a 25 Mw output has distinct cost advantages. More specifically, the range where the propane specific cost exceeds the minimum cost by less than 5% extends from 15 to 35 Mw. The ammonia turbines have a much broader optimum cost range extending from approximately 30 to 70 Mw.

Based on these results, the initial designs for 400 Mw powerplants, such as the Mark I and II systems, were designed with sixteen 25 Mw net power modules (approximately 30 Mw gross output) consisting of turbine, condenser and working fluid pump. Although it would have been desirable to have the same number of boiler modules, different constraints had to be applied to these components.

These included maximum height restrictions (it is desirable to keep a reasonably constant hot ocean water flow across the boilers) and maximum width restrictions (for hull design purposes). Application of these constraints yielded a Mark II design with fifteen separate boiler modules with a maximum system width less than 160 meters.

FIGURE 4.8: OPTIMUM SIZE FOR EACH TURBINE POWER UNIT

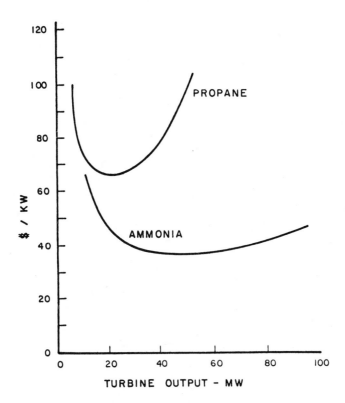

Source: PB 238 572

Site Location

The effect of site location was examined by consideration of temperature differences between surface and bottom water that are characteristic of tropical seas for three months of the year (high temperature, 85°F; low temperature, 40°F) and nine months of the year (high temperature, 82°F; low temperature, 45°F).

The design output for the two tropical plants using propane as a working fluid shows that the effect of greater available temperature differences between the working fluid and ocean water results in better heat transfer and subsequent reductions in heat exchanger core material volume. This fact is illustrated in comparison with a similar heat exchanger design, based on Gulf Stream conditions, in Table 4.4.

TABLE 4.4: EFFECT OF SITE CONDITIONS ON HEAT EXCHANGER VOLUME REQUIREMENTS (MARK II DESIGN, PROPANE)

Site Type	Temp Range, Hot–Cold, °F	Evaporator Material Requirements, 26.7 Mw (90/10 Cu-Ni), ft^3	Condenser Material Requirements, 25 Mw (90/10 Cu-Ni), ft^3	Total 400 Mw Powerplant Head Exchanger Material Requirements, ft^3
Gulf Stream	77–45	2,890	2,357	81,070
Tropical	82–45	688	2,374	48,300
Tropical	85–40	480	1,107	24,910

Source: PB 238 572

Heat Exchanger Fouling

An analysis was made of the effect of scale and fouling on the total heat exchanger material volume requirement by the inclusion of a unit conductance of scale in the expression for overall conductance of the heat exchanger. Results show an increase in heat exchanger material and consequently heat exchanger size as a result of the inhibition of heat transfer due to fouling.

Passage Size

Due to manufacturing considerations it is desirable to design the heat exchanger channel size as large as possible on the working fluid side, however, this causes heat exchanger material requirements to greatly increase. An investigation was made of the effect of working fluid channel size on total heat exchanger material volume for square channels ¼" on a side and ⅜" on a side.

The results of this preliminary study are shown in Table 4.5. Although these runs may not be the optimum (in terms of minimum exchanger volume) it can be seen that larger passage sizes require much more heat exchanger material and an increased core height, both of which make these alternatives unfeasible.

TABLE 4.5: EFFECT OF WORKING FLUID CHANNEL SIZE ON TOTAL MATERIAL REQUIREMENTS

Working Fluid Channel Size [in.]	Boiler Module Core Height [ft.]	Condenser Module Core Height [ft.]	Total Power Plant Material Requirement [ft.^{3}]
0.125 x 0.0938	29.9	28.3	81,070
0.250 x 0.250	119.5	108.7	146,300
0.375 x 0.375	114.3	82.8	2,394,000

Source: PB 238 572

Cladded Heat Exchangers

An investigation was made of the use of steel heat exchangers with ammonia as the working fluid which were clad with copper-nickel on the ocean water side. The steel exchangers were designed to be pressure-proof and the copper-nickel cladding was made 0.010" thick. Cladding achieved a 24% saving in the amount of copper-nickel used in the heat exchangers based on an ammonia baseline reference design.

TURBINE DESIGN (MARK I)

A Rankine cycle operating on the thermal differences of the Gulf Stream is very similar to a conventional fossil fuel powerplant. All of the major components are similar in function but different design criteria and operating conditions are required for each system. A schematic diagram of the closed cycle ocean thermal powerplant is shown in Figure 4.9. The most significant design condition imposed on this cycle is that of the low temperature differences of the ocean site.

In the case of the turbine this temperature difference or available head between states one and two is further reduced by heat exchanger losses. In considering an overall system analysis it is necessary to study carefully the tradeoffs between heat exchanger and turbine efficiencies. This interaction is particularly significant because of the fixed temperature differences. Power output can be written for the powerplant as follows:

$$P = \dot{m}\ \eta_p \eta\ H$$

In this η_p is the power efficiency of the system and is defined as the ratio of net power output to gross power generated. This term is a function of the parasitic losses in the system and therefore decreases as the mass flow rate of the working fluid increases. As a result a one percent decrease in turbine efficiency at a fixed power output would require significantly more than a one percent increase in mass flow rate.

Satisfactory plant performance is critically related to cycle and therefore turbine efficiency. It is therefore important to choose a turbine design that will give the highest practical efficiency. Relative to other components in the system turbine cost does not appear to be a major factor. Therefore, choosing turbine design for maximum efficiency is the design criteria.

Both axial and radial flow turbines could be used. Efficiencies above 90% have been obtained with both. When range of operation at high efficiency is compared on the basis of specific speed (turbine speed for unit head and unit flow), the axial machine has a wider range. Blade tip speeds are low enough so that stress considerations are not severe for either configuration. Radial flow machines are usually able to achieve higher stage loadings. For most fluids and the temperature differences, only a single stage is required for radial or axial flow.

Although both radial and axial flow machines appear acceptable at design point, some consideration must be given to off-design performance. For the Ocean Thermal Gradient (OTG) powerplant where variation in operating conditions are found during the year at each site, this is significant. Axial machines of fixed geometry operating at constant head have less variation in efficiency with

speed than a radial machine. Using partial admission for throttling again favors
axial machines. All considerations seem to favor the axial flow machine.

FIGURE 4.9: CLOSED CYCLE SEA THERMAL POWERPLANT SCHEMATIC

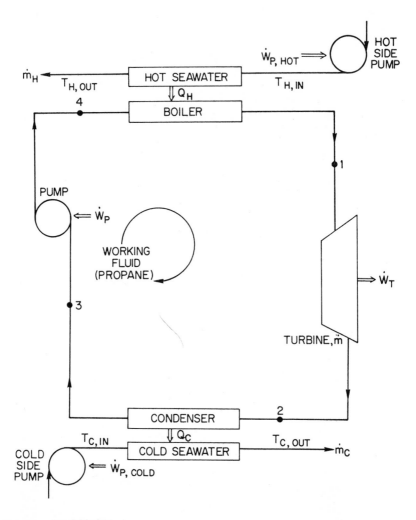

Source: PB 239 371

Working fluid selection is as important to the turbine as it is to other components in the cycle. However because of the restrictive cycle conditions of
an ocean thermal gradient powerplant the working fluid influence is somewhat
different from a conventional power cycle. Using a hot side temperature of
70°F and a cold side temperature of 40°F it is possible to determine cycle operating properties for working fluids. A summary for ten selected working fluids
is given in Table 4.6.

TABLE 4.6: WORKING FLUID CYCLE PROPERTIES

WORKING FLUID	VAPOR PRESS. AT 70°F	ΔP 70° to 40°	L BTU/lb	H BTU/lb	v_g 40°F	ṁ lb/sec for 100 mw	$MP_c L^{1.5}$	M	Turbine Size MW
Water	.3	.13	1071.	36.	2445.8	2,900	170,343	18	.15
Propane	124.3	82.8	156.7	6.	1.348	17500	7.85×10^6	44	3.1
Carbon Dioxide	853.4	199.8	95	2.9	.144	36400	26.6×10^6	44	4.72
Ammonia	128.8	39.6	536.2	20.2	3.971	5200	18.8×10^6	17	22
Sulfur Dioxide	49.1	16.2	162.2	6.1	3.02	17200	4.36×10^6	64.1	1.44
n-Butane	31.6	13.9	163.5	11.0	4.88	9600	1.99×10^6	58.1	3.9
Freon-12	61.4	23.4	64.1	2.5	.774	42100	2.36×10^6	120.9	.60
Freon C318	40.1	12.8	48.5	3.2	1.13	33000	2.06×10^6	200	.77
Freon 502	151.3	39.7	63.1	3.1	.447	34000	6.24×10^6	111.6	1.8
Genetrar 12/31	85.7	33.8	72.2	4.3	.907	24210	3.29×10^6	103.5	2.0

Source: PB 239 371

With the exception of water and carbon dioxide all have reasonable vapor pressures at the upper cycle working temperature. And although low pressure differential exists across the turbine for some of the candidate fluids all are practical. The isentropic enthalpy drop across the turbine does vary by almost one order of magnitude for the different fluids. Excluding water this leads to mass flow rates which differ by a factor of eight for any given power output. It would appear that ammonia and normal butane would be attractive fluids if flow losses are to be reduced.

Generally it is said that a high molecular weight is desirable for a working fluid so as to reduce turbine blade speed. This is usually useful to lower the number of turbine stages required. In this application the enthalpy drop across the turbine is so low that any of these fluids would only require one stage and in most cases the spouting velocity (isentropic velocity when expanded to zero pressure) is so low that speed induced blade stresses are not a problem.

Turbine size can be a problem however with OTG powerplants. With fluids having low isentropic head large flow rates are required for a given power output. For example a 100 Mw simple steam powerplant (800 psia, 900°F) would have a mass flow of about 250 lb/sec. All the candidate fluids in Table 4.6 have flow rates from 20 to 160 times greater. A grouping of terms which is a better indicator of relative size is also given in Table 4.6.

It can be shown from thermodynamic principles that the grouping (molecular weight x condenser pressure x latent heat$^{1.5}$) is proportional to the turbine output per unit area of the turbine exit. In order to reduce turbine size and cost, this parameter should be numerically large. If low flow rate and small turbine size are considered desirable, ammonia again appears to be the best fluid followed by normal butane and propane.

Using geometrically similar axial flow turbines operating under similar conditions, comparison of the fluids as to power output size for an individual unit shows that the ammonia unit size is larger than all other fluids by at least a factor of four. When building an overall OTG plant with 400 Mw net capacity ammonia would appear to be the best working fluid from a turbine point of view as it will require the fewest number of units. However for design studies, three working fluids were considered: ammonia, propane and refrigerant 12/31.

Results for ammonia, propane and R 12/31 at a fixed boiler temperature of 68°F and a series of condenser temperatures show that at a gross power output of 30 Mw per unit ammonia diameters range from 6 to 9.5 feet for condenser temperatures from 42° to 56°F. Turbine speeds run from 3,100 to 1,300 rpm for the same conditions. For a propane turbine size ranges from 8.5 to 14 feet and speeds from 1,200 to 475 rpm.

With R 12/31 as the working fluid the turbine size would range from 36 to 70 feet and the speeds from 240 to 70 rpm. This range of size and speed must be narrowed at the design point for satisfactory operation. At this power level the R 12/31 turbine size is prohibitive. If the same size R 12/31 turbine is used as was found for propane, the gross power output per unit would be about 1 Mw. R 12/31 seems to become an unreasonable turbine working fluid for these thermal differences.

With the use of ammonia and propane the optimum design point dimensions are strongly dependent on site conditions. Sixteen 30 to 35 Mw gross turbine units should produce the 400 Mw net capacity needed. Optimum design diameter and speed will depend on site thermal conditions as well as on heat exchanger design in the boiler and condenser. By fixing turbine power output and/or turbine speed, the range of operation can be determined as a function of specific speed and the other operating parameters.

At a fixed specific speed of 120 for a 30 Mw unit, the ammonia turbine can be run at 3,600 rpm with high efficiency if the temperature difference the turbine sees is 30°F or greater. For a 15°F temperature difference the turbine speed could be 1,800 rpm. For a 20 Mw propane turbine the temperature difference for high efficiency must exceed 30°F for 1,800 rpm speed. Increasing turbine speed is desirable because this reduces the required turbine size and cost.

Three turbine designs are compared in Table 4.7, one for each working fluid. All are for operation with the same internal temperature difference of 16°F.

TABLE 4.7: TURBINE DESIGN COMPARSION (T_B = 68°F T_C = 52°F)

Ammonia	Propane	R 12/31
P = 30 mw	P = 20 mw	P = 20 mw
D = 8 ft.	D = 9.6 ft.	D = 52 ft.
N = 1800 RPM	N = 820 RPM	N = 105 RPM
ṁ = 1950 lb/sec	ṁ = 4400 lb/sec	·ṁ = 9000 lb/sec

Source: PB 239 371

The actual size of the turbine used for any of the fluids depends upon the site conditions and the heat exchanger design. Based on the fluid mechanical and thermodynamic performance of the turbine design, ammonia is the preferred working fluid. Although requiring a somewhat larger and slower unit of less power, propane would also be acceptable.

TECHNICAL AND ECONOMIC EVALUATION OF TURBOMACHINERY (MARK I)

Turbine Size

The ocean thermal difference (OTD) cycle is inherently inefficient because of the low temperature difference over which the cycle operates. Consequently, it is vital that each component operate as close to maximum efficiency as is economically possible.

From a study done on turbine performance, it would appear that it would be possible to have a turbine total-to-static efficiency greater than 90%. The study revealed that an efficiency of approximately 91.4% is possible in the specific speed (N_S) range of 100 to 150. (See Table 4.8 for definition of N_S.)

TABLE 4.8: DEFINITION OF N_S AND D_S

Specific Speed (N_S)

$$N_S = \frac{N\ V^{\frac{1}{2}}}{H^{3/4}}$$

N = rotative speed, rpm

V = volume flow at turbine exit, ft^3/sec

H = adiabatic head, ft

Specific Diameter (D_S)

$$D_S = \frac{D\ H^{\frac{1}{4}}}{V^{\frac{1}{2}}}$$

D = rotor diameter, ft

H = same as above

V = same as above

Output (P)

$$P = \frac{\eta p}{.738\ RT}\left(\frac{N_S}{N}\right)^2 H^{5/2}$$

P = turbine output, Mw

η = turbine efficiency

p = static pressure at turbine exit, lb/ft^2

R = gas constant, ft/R

T = static temperature at turbine exit, R

N_S = same as above

N =   "

H =   "

Source: PB 239 373

The study also indicated that both the ratio of exhaust energy to available energy, and the blade height-to-diameter ratio are greater at a specific speed of 150 as compared to 100.

It is desirable to have as low an exhaust energy as possible to minimize the problems of physically incorporating a diffuser in the powerplant. Likewise, it is desirable to minimize blade length to reduce the stress level at the root of the blade.

Consequently, a specific speed which is as low as possible, but which would also maintain high turbine efficiency, was selected, i.e., N_S = 100. The optimum specific diameter (D_S), which corresponds to a N_S = 100 is 1.3. (See Table 4.8 for definition of D_S.) Thus, the two basic parameters for establishing the turbine size have been evaluated: N_S = 100, and D_S = 1.3.

Figures 4.10 and 4.11 display rotational speed, rotor tip diameter and blade height as a function of turbine output power. Figure 4.10 is for the ammonia working fluid while Figure 4.11 is for propane. In both illustrations, it can be observed that increasing the output power increases the rotor tip diameter and decreases the rotational speed. By comparing the two illustrations, it can be further seen that, for a given turbine output power, the propane turbines are larger in diameter and slower in rotational speed as compared to the ammonia turbines.

Tables 4.9 and 4.10 and Figure 4.12 summarize the remaining technical parameters related to the turbine which have been evaluated. The tables cover a range of possible rotational speeds and the corresponding turbine output for each working fluid. The geometric parameters defining the fluid flow for the nozzle and vanes are illustrated in Figure 4.12.

These results are generally self-explanatory; however, there are several comments which could be made. The results for the value where the ratios of blade tip clearance to blade height down to 0.02 were considered are indicated in Tables 4.9 and 4.10. However, the blade clearances are quite large in light of the small strain induced in the blades. It is conceivable that the clearance ratio could be even smaller, say 0.0005. An efficiency of 91.4% was used.

The results indicate a 58% reaction turbine stage. A 50% reaction turbine stage would result in symmetrical vanes and nozzles, thus reducing manufacturing costs. Unfortunately, it would appear that any attempt to reduce the degree of reaction would result in an unacceptable excursion from maximum efficiency for the turbine.

Neither the ammonia or propane turbines appear to be constrained in their design by blade root stress considerations. The stress resulting from steel blading was indicated in Tables 4.9 and 4.10. However, the centrifugal force involved is low enough so that the stress level induced is well within the level which can be borne by many other substances including plastics and wood. The magnitude of the tip speeds is such that, neither the tip speeds (U) nor relative velocities (W) are supersonic. This is desirable since Mach numbers greater than 1.25 could result in considerable efficiency degradation.

FIGURE 4.10: AMMONIA TURBINE CHARACTERISTICS

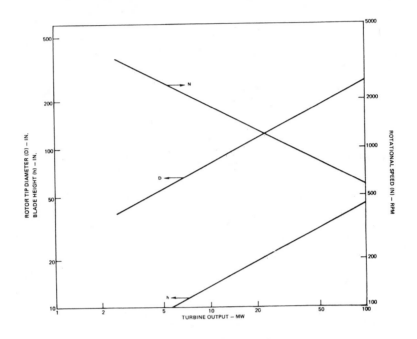

FIGURE 4.11: PROPANE TURBINE CHARACTERISTICS

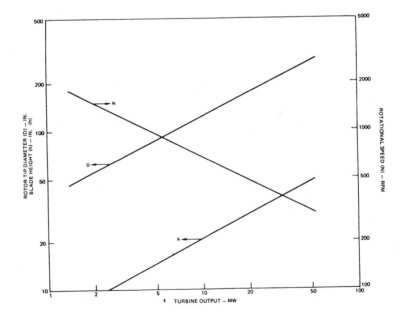

Source: PB 239 373

TABLE 4.9: AMMONIA TURBINE CHARACTERISTICS

$N_S = 100$
$D_S = 1.3$
$\eta = .914$

Parameter	Rotational Speed, rpm			
	1,800	1,200	900	600
Turbine Output, Mw	10.79	24.29	43.18	97.18
Tip Diameter, in.	83.5	125.2	166.9	250.4
Blade Height, in.	14.2	21.3	28.4	42.6
Blade Chord, in.	23.4	35.1	46.8	70.1
Blade Axial Width, in.	7.9	11.8	15.7	23.6
Blade Tip Clearance, in.	0.29	0.43	0.57	0.85
Blade Root Stress, psi	13,055	———→		
Nozzle Blades	20	———→		
Rotor Blades	14	———→		
Tip Velocity, ft/sec	665	———→		
Sonic Velocity, ft/sec	1,415	———→		
Ratio of Exhaust Energy to Available Energy	0.025	———→		

TABLE 4.10: PROPANE TURBINE CHARACTERISTICS

$N_S = 100$
$D_S = 1.3$
$\eta = .914$

Parameter	Rotational Speed, rpm			
	900	600	400	300
Turbine Output, Mw	5.81	13.48	29.44	52.36
Tip Diameter, in.	91.9	137.9	206.9	275.9
Blade Height, in.	15.6	23.4	35.2	46.9
Blade Chord, in.	25.7	38.6	57.9	77.3
Blade Axial Width, in.	8.6	13.0	19.5	25.9
Blade Tip Clearance, in.	0.31	0.47	0.70	0.94
Blade Root Stress, psi	3,963	———→		
Nozzle Blades	20	———→		
Rotor Blades	14	———→		
Tip Velocity, ft/sec	361	———→		
Sonic Velocity, ft/sec	832	———→		
Ratio of Exhaust Energy to Available Energy	0.025	———→		

Source: PB 239 373

Thermal Energy from the Sea

FIGURE 4.12: TURBINE GEOMETRIC PARAMETERS

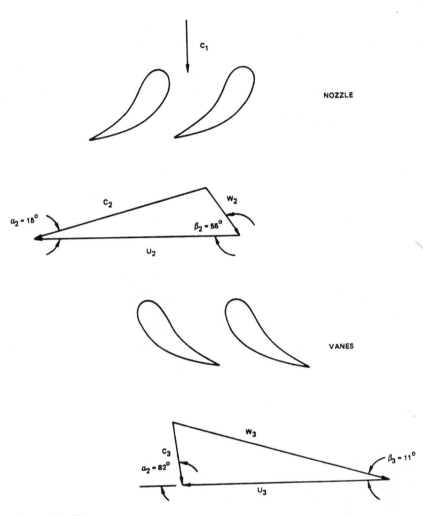

Source: PB 239 373

Turbine Cost

The cost estimates are shown in Figure 4.13. The cost presented in this illustration is for a single turbine and includes manufacturing, assembly, continuing engineering, overhead, profit to the manufacturer, and an allowance for auxiliaries such as bearing coolant pumps and controls. It does not include the development cost.

FIGURE 4.13: COST OF INDIVIDUAL TURBINE

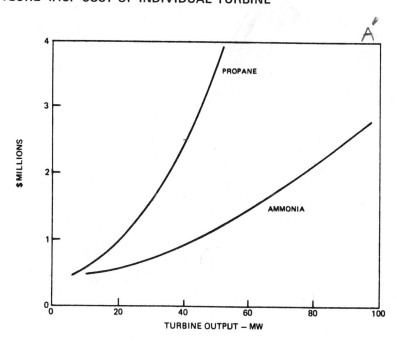

Source: PB 239 373

In addition, it does not include the transportation cost, since the prices are F.O.B. factory, nor does it include the installation cost. Installation costs have historically amounted to about 30% of equipment costs for typical installation. This is not a typical installation.

In Figure 4.13, it can be observed that propane turbines are considerably more expensive than ammonia turbines at any turbine power output level. This fact is stated in Figure 4.14 where the preceding data has been converted in specific cost and replotted as a function of turbine output. The minimum propane turbine cost of $47/kw is more than twice the minimum ammonia turbine cost of $22/kw.

FIGURE 4.14: SPECIFIC COST OF INDIVIDUAL TURBINE

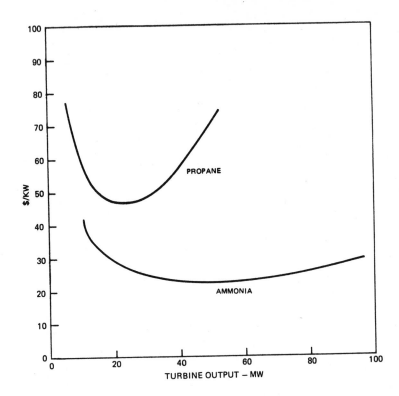

Source: PB 239 373

It should be noted that the ammonia turbines have a large range of output wherein the specific cost is less than 5% greater than the minimum specific cost. This range extends from about 30 to 70 Mw. Conversely, the propane turbine specific cost is more sensitive to turbine output. The range wherein the specific cost exceeds the minimum cost by less than 5% extends from about 15 to 35 Mw.

The total turbine output, pump inputs and powerplant net output are presented in Table 4.11. It is obvious from the preceding discussion that the total turbine output can most economically be achieved by incorporating groupings of individual turbines. It is equally obvious that the smaller the individual turbine output, the greater the number of turbines required. Table 4.12 indicates, for a specific turbine size and rpm, the number of turbines needed, and the total cost for all the turbines.

Perhaps a more meaningful comparison is presented in Figure 4.15 where a specific cost of all the turbines is plotted as a function of individual turbine output. The values of net powerplant output, as presented in Table 4.11, were used as the denominator for dividing the total cost in order to obtain system specific cost. Similar characteristics which were exhibited for the individual turbines have occurred again for the system. A minimum value of $60/kw at 20 Mw turbine output was obtained for propane while a value of $36/kw was obtained for ammonia at 45 Mw.

FIGURE 4.15: SPECIFIC COST OF TURBINE SYSTEM

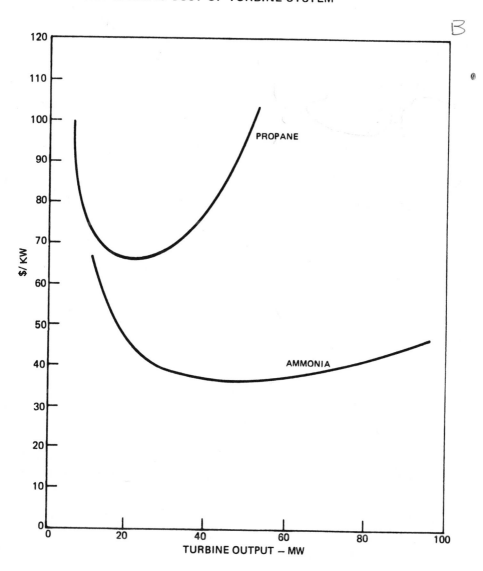

Source: PB 239 373

Thermal Energy from the Sea

TABLE 4.11: OTD POWERPLANT POWER PROFILE

	Mw	
	Ammonia	Propane
Total Turbine Output (η = .914)	651.7	437.8
Cycle Pump Input	8.74	21.80
Cold Water Pump Input	168.19	67.15
Hot Water Pump Input	64.68	6.16
Net Output	410.1	342.7
Cycle Thermal Efficiency (%)	1.31	1.64

TABLE 4.12: TOTAL TURBINE COST

Ammonia

RPM	Turbine Output, Mw	Number Needed	Unit Cost $ x 10^{-3}	Total Cost $ x 10^{-6}
1800	10.79	60.4(61)	452.7	27.61
1200	24.29	26.8(27)	638.3	17.23
900	43.18	15.1(16)	998.5	15.97
600	97.18	6.7(7)	2772.7	19.41

Propane

RPM	Turbine Output, Mw	Number Needed	Unit Cost $ x 10^{-3}	Total Cost $ x 10^{-6}
900	5.81	75.3(76)	452.2	34.36
600	13.48	33.4(34)	689.0	23.43
400	29.44	14.9(15)	1573.0	23.60
300	52.36	8.4(9)	3423.1	30.81

Source: PB 239 373

Generator Cost

The cost of the generator is presented in Figures 4.16 and 4.17. Figure 4.16 presents the specific cost for an AC synchronous generator while Figure 4.17 presents the cost for a DC generator up to 10 Mw in size. These prices are several years old and should be increased by about 15% to bring them up to 1973. These prices are F.O.B. selling prices and do include development which has been amortized over a large number of units. They do not include the approximately 15% premium required for auxiliaries and controls, transportation, and the 30% required for installation. In the range of outputs being considered for ocean thermal difference powerplants, it appears that an AC synchronous generator, of U.S. manufacture, would cost between $17 to $20/kw.

System Losses

Perhaps the most significant area unaccounted for in the present analysis results from the cumulative effect of the inefficiencies in the internal powerplant operation. For example, it is presently proposed to use a turbine to drive a generator which, in turn, would power the motors used for internal pumping.

In this sequence of events, only the thermodynamic efficiencies have been accounted for in this analysis. The bearing losses in the turbine and in the pumps, the conversion losses in the generator and in the motor, plus the transmission losses from the powerplant to land have not thus far been adequately considered.

A preliminary estimation was performed utilizing the ammonia turbine to determine the magnitude of the power losses which are involved. The results revealed losses amounting to 13% of the expected net output. This loss increases the specific cost of the powerplant and all its components, including the turbines and generators, by about 15%.

This exploratory analysis did not consider the losses which would be involved in either converting the electric energy to a storable form on site, such as hydrogen, or transporting the electric energy to the load center. The magnitudes involved merit an in-depth look at these losses.

TECHNICAL AND ECONOMIC FEASIBILITY (MARK II)

The machinery concept chosen for Mark I is that of a turbine inside a one-atmosphere machinery compartment, manned for maintenance and repairs, the turbine sitting on top of a condenser which can be isolated from sea pressure for maintenance and repairs, and topside evaporators fully immersed in the oncoming hot surface water.

Maintenance and repair of evaporators will be accomplishable by surfacing the powerplant (at the cost of total stoppage in production). There is thus considerable redundancy against either planned or unexpected outage of turbines or condensers, but poor redundancy against mandatory evaporator repair.

FIGURE 4.16: AC SYNCHRONOUS GENERATOR SPECIFIC PRICES (1968 Dollars)

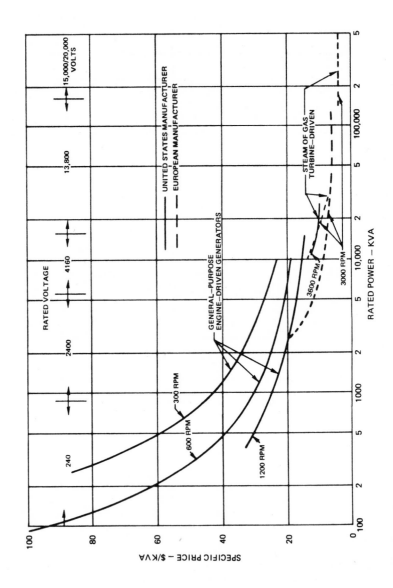

Source: PB 239 373

FIGURE 4.17: DC GENERATOR SPECIFIC PRICES (1968 Dollars)

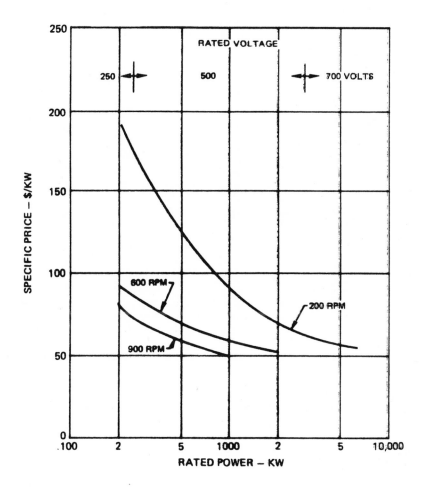

Source: PB 239 373

It was originally planned that the machinery compartments would be manned and watchstanders would stand watch and live in appropriate spaces down inside the hulls, but it is now planned to move all personnel topside, above windwaves, remoting all data logging and controls, except for maintenance and repair visits down inside the hulls.

The same "habitable machinery compartment" philosophy has been retained in the Mark II baseline—nothing has been done yet about a completely submersible unmanned powerplant of the type proposed by other investigators.

The machinery power package comprises evaporators, turbine, condenser, feed pumps, cold water circulating water pumps, hot water circulating water pumps, working fluid management subsystems, and other machinery plant subsystems. The total thermal cycle including all parasitic losses has been modeled—the model sizes and describes the individual components.

The turbine has been established as "king" to insure that its efficiency will be as high as practically feasible, then all other subsystems evolve thereafter from the thermal cycle program. Constraints reflecting arrangement problems have been introduced into the thermal cycle analysis. The major control, however, is still economics (i.e., amount of material in the heat exchangers per installed kw).

Propane has been selected as the working fluid for the Mark I study temporarily setting aside the ammonia cycle. Perfection of the plate-fin heat exchanger geometry has been attempted, setting aside temporarily staggered tubes.

The size of the cold water inlet pipe has been held constant while investigating variations in that major subsystem were investigated. Once the variation in cost of the CWIP as a function of its hydraulic radius has been determined, reduced size CWIP versus enlarged heat exchangers can be traded off.

Pump work on the hot water side in all of the thermal cycle studies has been included except for the specific study which showed that the Gulf Stream momentum can do the hot water pumping if powerplant size is held much lower than 400 Mw_e (within certain physical constraints tied to anchor capability). That study must be carried to the conclusion of an orderly trade-off against the water-pumping competitor.

It has been concluded that break-down in thermal stratification will not occur at Site One using the Mark I CWIP configuration and flow rate. It has been decided that withdrawal of cold water should best start off with a $90°$ turn from horizontal flow into vertical flow, with a horizontal baffle plate at the base of the CWIP.

Progress has been made toward characterizing year-round hot and cold water conditions at other sites, and progress has been made toward preparation of smoothed annual plots of T_{hot} and T_{cold} for Site One, basic inputs into the off-design point thermal cycle analyses.

The turbine investigation has produced a complete machinery arrangement from which vapor headers can now be fitted into the machinery compartments and their necessary hull penetrations considered. Working fluid storage and management subsystems, condensate and boiler feed systems are now shaped up in real sizes, and subsystem descriptions are improving.

The complete turbine has been integrated into the thermal cycle model, taking the place of the "black box, $\eta = 90\%$" that had been there earlier. Steady progress toward the ability to manage the off-design point analyses has been made. It has been agreed that the turbine can be relegated to the second tier of difficulty; one could contract for the detail design and fabrication of the propane turbine at any time.

The structural mechanics and design work shifted to an analysis of detail in the proposed steel reinforced concrete hulls. One early proposed Mark II structural configuration which put the circulating water aqueducts inside the cylindrical shell, and used a continuous high pressure tank top (deck) over the condenser boxes, has been analyzed in great detail: those results call for shell walls thicker than the 24" with which Mark I started.

This result has led to other structural configurations which appear to have smaller pressure hull diameters and easier details, but that remains to be seen. The estimated cost for steel reinforced concrete, in place, for the kind of very large shipyard type fabrication being contemplated, is much higher than 1972 estimates ($350/yd^3 vice $100/yd^3) and concrete hull costs have now grown to be significant.

There will therefore be considerable drive toward reduction in hull size and hull simplification. Since those parameters are set by condenser size and the hydraulic losses in the cold water circulating system, the trade-off is a head-on collision.

The latest smaller hull concepts for the Mark II baseline contain large openings in a cylindrical shell. The practicality of supporting those openings with structural gratings (of streamlined detail) through which the circulating water may pass is being investigated.

The naval architecture continues to be the tedious analysis of static balance and the slowly-changing trimming problem plus the dynamics of this huge craft on the way to the site and in place at the site. This latter investigation is now very nicely integrated with the dynamic analysis of the CWIP.

Results to date are reassuring, but much more work is required before there is the required confidence about loadings expected on this system, in a seaway, on the way to the operating site.

Attention has been given to the feasibility of construction of the powerplant in an existing shipyard, but to nowhere near the detail appropriate. It is possible that the individual hulls (two of them) and the CWIP will have to be constructed as entities, launched, then joined together while water borne in a rather large outfitting basin.

Some progress has been made toward establishing a position regarding the biofouling problem. No one challenges the idea that biofouling would probably be no problem at all on heat exchangers if high copper alloy (90/10 Cu-Ni alloy) were used as exchanger material. There will be the same kind of microorganism film that forms momentarily on copper-nickel condenser tubes but which seems to be sloughed off periodically.

The mechanisms by which Cu-Ni keeps itself bright in seawater would do the job for this powerplant. Most seem to agree that moving water with a 2.5 fps or higher velocity will tend to minimize fouling. There is general concurrence that if a 4.0 fps or higher velocity were maintained that no fouling of the large size creatures would be expected on any surface.

There seems to be no real knowledge as to what fouling creatures may be expected out in the upper layers of the flowing Gulf Stream. Any real investigation of antifouling measures should start, then, with an assessment of actual fouling populations at the intended sites. It is thought by some that there may actually be very few fouling creatures out there in those warm but rather nutrient-deficient waters.

At the other end of the circuit, there seems to be a consensus that very few if any fouling creatures will be drawn in from the cold bottom water so long as scouring of the seabed is avoided. Again, the real investigation of antifouling measures required in the cold water circuit should start from an assessment of actual populations.

It has been thought that a metal other than high copper alloy or a plastic might suffice in the cold water circuit because of low occurrence of fouling creatures. At least one opinion has been advanced that essentially no microorganism build-up on the cold water surfaces can be expected. Groove type enhancement on the seawater side of the condenser might therefore stay clean and effective.

It is clear that on site generated chlorine can be used to prevent fouling if necessary, but the power drain will probably be much greater than calculated a year ago. Commercial equipment could be purchased to accomplish the chlorination if that path is selected.

Regarding the question of need for fouling control on exterior hull surfaces, it was suggested that accumulating fouling would certainly increase drag and thus load on the mooring even if it didn't throw the system out of trim. The reinforced concrete hulls do not want to be coated with a water-tight paint or antifouling film: it may be that copper-nickel mesh fastened to such surfaces could restrict fouling and still permit continuous hydration of concrete surfaces. For the enclosures around the evaporators and similar high structure, antifouling paint or thin section "coppering" could be used.

COMPARATIVE STUDY
OF CLOSED CYCLE TECHNOLOGIES

The information in this chapter is excerpted from the following publication:

NASA TM X 70783

COST COMPARISON OF CMU AND UMASS DESIGNS

Carnegie-Mellon University (CMU) SSPP is a 100 megawatt plant. The cold water pipe for this plant will be 2,000 ft long and 60 ft in diameter and bring 10 million gal/min up to the SSPP condensers. The CMU plan also calls for 8 ft/sec water flow through the heat exchanger tubes. These tubes will be 10 ft long and 1 inch in diameter.

The thermal gradient that will operate this SSPP is assumed to be 20°C of which 10°C will be dissipated in forcing the heat to the working fluid. This leaves 10°C available for the heat engine and yields an efficiency of a bit over 2%. The weight of the tethering cable is estimated to be 800,000 lbs. Table 5.1 compares the cost of the CMU SSPP with the UMASS concept described by Anderson. The life of the CMU SSPP is assumed to be 25 years with maintenance and the heat exchangers are assumed to be the life limiting components.

The UMASS SSPP, designed to be placed in the Gulf Stream 15 miles east of the Collier Building of the University of Miami, assumes a thermal working gradient of 31.5°F and an efficiency of 1.8%. Propane is to be the working fluid for this 400 megawatt SSPP. The cold water pipe will be fabricated out of aluminum or mild steel and the plate and fin heat exchangers will be made of CuNi.

The UMASS SSPP is to have 16 power units and 20 vertical heat exchangers. A 40 year life is assumed with a cost of $300/kw to $750/kw delivered at the beach. The turbine will cost $65/kw if propane is selected as the working fluid, but only $35/kw if the working fluid is ammonia.

TABLE 5.1: 100 MEGAWATT SSPP

	UMASS Thousands $	CMU Thousands $
Boiler evaporator	2,400	1,600
Turbines	720	1,000
Generator	1,235	1,800
Boiler water pump	820	1,000
Condenser	2,450	1,600
Cond. water pump	640	1,000
Warm water pipe	131	250
Cold water pipe	708	1,000
Inlet screens	304	450
Auxiliary pumps	152	230
	9,560	9,930
Auxiliaries	443	650
Structure	4,210	6,000
Cold pipe	262	400
	14,475	16,980
Miscellaneous	2,172	3,020
	16,647	20,000

HEAT EXCHANGERS

The greatest economic problem in the construction of a closed cycle SSPP is the construction of low cost heat exchangers. The greatest technological problem in closed cycle SSPP design is to design and build heat exchangers that transmit approximately 30 times as much heat per kilowatt as is now being done by the heat exchangers in typical fossil fuel power plants. Moreover this feat must be performed while making use of much smaller temperature differences than those which conventional power plants make use of in their operation. The small ΔT and the need for very high efficiency dictate to the SSPP designer that he must design massively large heat exchangers. Since the heat exchangers will necessarily be large a way must be found to keep the cost down and then insure long lifetimes for the heat exchangers so as to make the SSPP investment attractive.

The Carnegie-Mellon University heat exchanger design consists of hexagonally nested 10 ft pipes, one inch in diameter, through which the tepid surface water is pumped and the working fluid, which flows around the pipes, is consequently boiled. On the water side of the tubes, the tepid water flows with a velocity of 8 ft/sec. The temperature losses that are intrinsic properties of this heat exchanger design are the following: a loss of $4°C$ through the 3 mil film, a $2°C$ loss through the aluminum tube wall and another $4°C$ loss through the 3 mil film on the working fluid side with working fluid having been imparted a velocity of 8 ft/sec. This adds up to a total loss of $10°C$ due to the heat exchanger operation. Since one has only a ΔT of the order of $20°C$, one half of the available ΔT is dissipated by the heat exchanger's operation.

The CMU design also incorporates 25 mil depth flutes on the outside, or working fluid side of the heat exchanger pipes. It should be noted that flutes on the tepid water side of the heat exchanger or on the cold water side of the heat exchanger (condenser) would create near optimum sites for biological growth due to the tendency of the flutes to induce local turbulence.

The UMASS SSPP heat exchanger is of the plate and fin type make of CuNi in a 90 to 10 ratio. Plastic was also considered as a heat exchanger material because of its low cost, as was aluminum. CuNi was chosen in spite of its high cost because of its alleged anti-biofouling properties.

The UMASS study indicates that compact heat exchangers of a plate-fin design with extended interior exchange surfaces are superior to plain tube and shell exchangers. The second result is that they could not determine the optimum plate-fin heat exchanger design. Small diameter tubes appear to offer advantages over larger tubes, however, one must take the possibility of clogging into consideration and also the difficulty involved in manufacturing small diameter tubes. The UMASS study indicates that none of the current enhancement techniques will succeed in raising the heat transfer efficiency over that obtained with internally extended surfaces.

Biofouling: The UMASS system considers the problem of biofouling of the heat exchanger external surfaces serious. Recognition is given to the fact that exposing the heat exchanger surfaces to sea water causes those surfaces to become covered by a thin film of microbial growth that will essentially greatly reduce the efficiency of the boiler and condenser. The results of this study indicate that one part in four million of chlorine is sufficient to prevent microbial growth. It is doubtful that these results would apply to the heat exchanger situation in an SSPP with its massive flows of water and continuously distributed points of localized turbulence. Biofouling thus remains an unresolved problem. Three basic biofouling prevention methods (to be applied in light or dark situations) are presented in the UMASS study.

Move the tepid water past the boilers with a velocity greater than 2.2 knots.
Use CuNi to construct the heat exchangers to prevent biofouling.
Continuously wet the heat exchanger surfaces with a sodium hypochlorite solution.

Both UMASS and CMU studies seem to feel that the biofouling problem can be easily solved and it is only the economics of the solution that gives any cause for concern. Little concern is given to the fact that one must pump on the order of 5,000,000 gal/min of tepid water past the boilers and about twice that, or 10,000,000 gal/min, of nutritious cold bottom water past the condensers which will warm it up a bit. Moreover, the creation of an artificial upwelling zone surrounding the SSPP implies that even the tepid water will be saturated with nutrients conducive to biological growth and, indeed, this water will have an increased number of organisms that are potentially detrimental to SSPP heat exchanger heat conduction properties.

It is evident that biofouling is the most serious problem associated with heat exchanger design especially because of the small ΔT available and the high efficiency that the SSPP must operate at to be classified as successful. The projected lifetimes of the SSPP heat exchangers are 25 years for the CMU design and 40 years for the UMASS design. None of the methods proposed for bio-fouling prevention can be successful in an open system for any length of time.

Corrosion, Stress Corrosion, Corrosion Fatigue and Material Aging: Corrosion, stress corrosion and to a lesser extent corrosion fatigue and material aging are likely to present some difficulties in the design of the heat exchangers. The corrosion problems are not serious in that one can design corrosion resistant structures and components. In the heat exchanger situation one must optimize heat transfer characteristics in conjunction with optimizing anticorrosion prop-erties. If this optimization is performed with cost as a constraint it may turn out to be a somewhat difficult chore with some difficult trade-offs taking the designed lifetime of the SSPP as another constraint.

In addition to the choice of material there are other means available to augment heat transfer. In one technique, spots of Teflon or other nonwetting material on the heated surface promote nucleations. There are many other techniques to augment heat transfer, however, most use up more energy than they add by their augmentation.

An interesting alternative SSPP heat exchanger has been proposed. This alterna-tive involves applying the concepts of heat pipes and thermosyphons in the heat exchanger design. Heat pipes and thermosyphons are both scaled objects, com-monly cylindrically shaped, containing a fluid present predominately as a vapor. This fluid is at the saturation temperature corresponding to the vapor pressure.

The thermosyphon is a vertical device that transfers heat from below the device, boiling the fluid and raising the internal pressure and temperature. To move towards equilibrium heat must be transferred from the top portion of the ther-mosyphon. This condenses the vapor at the top and the liquid returns under gravity to the evaporation portion of the thermosyphon.

The heat pipe is a thermosyphon wherein the condensate flows under the action of capillary forces, rather than under the action of gravitational forces as in the thermosyphon. In designing an SSPP heat exchanger one may make use of both gravity and capillarity.

In practice the heat pipe is simply a tube or duct whose wall is clad with a layer of porous material, usually called the wick. Since both evaporation and con-densation can occur at almost the same temperature, the total temperature drop over a heat pipe may be very small, a few degrees or so, as in SSPP applications, whereas the heat transported may be of the order of kilowatts.

The effective thermal conductance can, therefore, be more than tens of thou-sands of times that of a copper rod of the same size. Add the assistance of gravity and one has a truly marvelous heat transfer device.

In the SSPP heat exchanger situation the heat-transport capability is moderate and perhaps it offers no advantage over other heat exchanger suggestions. The heat transport capability of a heat pipe is due in large part to the fact that the thermal resistance of the vapor duct is very small. The heat flux is limited by the heat transfer into and out of the pipe and the evaporation and condensation

TABLE 5.2: EVALUATION OF SSPPs USING VARIOUS WORKING FLUIDS

Working Fluid Items	R11	R12	R13	R21	R22	R113	R114	NH$_3$	C$_2$H$_6$	C$_3$H$_8$	C$_2$H$_4$	Water
Adiabatic Enthalpy Drop	2 Kcal/kg	2.2	1.1	3.5	1.9	2.8	3.1	8.0	3.0	5.0	2.0	7.2
Turbine Inlet Flow	56.4 m^3/s	80.3	2 39	21.5	8.5	5.6	11.5	13.7	2.63	6.02	1.51	3.760
Turbine Inlet Pressure	0.985 ata	6.19	32.3	2.0	9.5	0.59	2.4	7.59	39.0	9.63	67.9	0.0363
Turbine Exhaust Pressure	0.557	4.08	23.9	1.10	6.51	0.25	1.3	58.0	30.3	0.12	51.0	0.0125
Turbine Blade Height	0.516 m	0.542	0.086	0.214	0.177	0.548	0.179	0.144	0.064	0.101	0.046	7.7
Mean Diameter	1.64 m	1.93	0.5	1.24	0.66	1.62	0.855	0.7	0.43	0.55	0.35	5.2
Boss Rates	0.6	0.6	0.78	0.77	0.66	0.59	0.723	0.72	0.8	0.75	0.835	0.5
Weak Points	Too large mean diameter Too long blade length Too large volume flow	Too low enthalpy drop Too high pressure	Too low enthalpy drop Too high pressure	Mean diameter large	Pressure somewhat high	Too large mean diameter Too large volume flow Too long blade length		Pressure somewhat high	High pressure	Pressure somewhat high	Too high pressure	Too large mean diameter too low pressure Too large volume flow
Probability of Realization	△	△	△	○	○	△	◎	○	△	○	△	✕

◎ Realizable   ○ Probable   △ Hard   ✕ impossible

processes. One advantage of the heat pipe is that it is a reversible unit. Thermal energy may be transferred through heat pipes from either section to the other section. Thus one may use ammonia as a cleaning (and toxic) agent as well as a working fluid simply by mechanically reversing the heat exchangers. This type of design is likely to be expensive and seals will have to be designed. In addition the use of heat pipes or thermosyphons result in the construction of larger heat exchanger units.

WORKING FLUIDS

The major contenders for the working fluid job are: (a) ammonia, (CMU); (b) propane, (UMASS); (c) the halo-carbon R-12/31; and (d) R-114, (see Table 5.2). All of candidates yield cycle efficiencies near the same theoretical maximum. Table 5.3 illustrates some of the UMASS results on working fluids.

The working fluid should be optimized with respect to the heat exchangers and not the turbines, as the heat exchangers are the critical components in the SSPP design. Turbine operation for (almost) any working fluid is easily optimized.

With respect to the environmental effects in the event of a massive working fluid spill, ammonia will have the smallest environmental impact as it is easily assimilated and dissipated in the ocean. The other three are known to be inferior to ammonia in this respect, but their precise environmental effects must be determined in advance if they are to be used as working fluids.

TABLE 5.3: COMPARISON OF WORKING FLUIDS FOR IDEAL 100-mm CYCLE

Fluid	Cycle Efficiency (%)	High Pressure (psia)	Low Pressure (psia)	Pump Work (kw)	Mass Flow (lb/min)
Ammonia	3.72	118	81	1,079	317,600
Butane	3.82	29	20	859	976,000
Carbon dioxide	2.89	779	609	36,003	2,873,000
Ethane	3.90	411	53	25,300	1,495,000
R-12	3.68	78	56	2,450	2,630,000
R-12/31	3.71	79	57	2,150	2,200,000
R-22	3.68	126	91	3,200	1,978,000
R-113	3.65	5	3	170	2,436,000
R-502	3.61	140	103	4,552	2,756,000
Propane	3.67	115	85	3,706	1,084,000
Sulfur dioxide	3.72	45	30	634	1,041,000
Water	3.78	0.3	0.15	1.4	155,500

TURBINES

If ammonia is selected the major turbine problems become the turbine seals and turbine materials. In fact no matter what the working fluid selected, the SSPP cannot tolerate working fluid leakage by way of the turbine shaft seal. The crucial material problem centers on the question as to whether or not turbine materials can be developed that will not deteriorate no matter how wet the ammonia becomes.

The UMASS turbine, selected with the constraint that the working fluid is propane, will generate 30 megawatts per unit (Table 5.4). The cost is estimated at $65/kw. However, if ammonia is chosen the cost decreases to $35/kw.

TABLE 5.4: OPTIMUM PARAMETERS FOR TURBINES WITH AMMONIA AND PROPANE AS WORKING FLUIDS

(A) Ammonia Axial Turbine (100 megawatt total output)

    1. Specific speed = 100. Specific diameter = 1.0
    2. Use three units of 33.3 megawatts each
    3. Single storage, full periphery entry
    4. Wheel diameter approximately 7 feet
    5. Wheel rpm about 1800
    6. Cost: $35/kw

(B) Propane Axial Turbine (100 megawatt total output)

    1. Use three units of about 38 megawatts each
    2. Single stage, full periphery entry
    3. Wheel diameter approximately 11.4 feet
    4. Wheel rpm about 600
    5. Cost: $65/kw

COLD WATER SUPPLY AND PIPE

The proposals for pipe include: a welded aluminum pipe aerodynamically shaped where the pipe for a 400 megawatt plant would be teardrop shaped and be 80 ft by 225 ft; for a 100 megawatt plant, a 40 ft diameter pipe made of vertical 8 inch piping; a 60 ft diameter pipe for a 100 megawatt plant; and an FRP pipe called Takata HF pipe proposed in a Japanese study. This pipe can be shipped folded flat and then inflated and cured by air and heat at the SSPP construction site. The upper portion of this pipe should be reinforced near the top to protect the pipe against wave damage. In any case the pipe should be designed keeping in mind that certain potential problems may arise. The problems are:

    (a) Vortex induced loading and the pipe's response
    (b) Internal fluid flow and pipe instability
    (c) The coupling of (a) and (b)
    (d) Water hammer effects
    (e) Response of pipe to wave induced motion of SSPP
    (f) Fatigue and corrosion fatigue associated with the loading generated in (a) - (e) above.

For streamlined cross section pipe local subsurface currents (flow directions are not always constant with depth) that would tend to apply bending moments to the cold water pipe should be considered. A small diameter 870 meters deep cold water pipe made of polyethylene has been reported to operate successfully.

MOORING AND ANCHORING

The UMASS SSPP at the Miami site has been calculated to be about 16,000,000 lb. The moorings needed by SSPPs must be easily deployable, have lifetimes of 25 to 40 years, be capable of being repaired easily and have almost 100% reliability.

ADDITIONAL TECHNOLOGIES

The information in this chapter is excerpted from the following publications:

NASA TM X 70783
NASA SP 5110
U.S. Patent 3,403,238
PB 228 066

HYDROGEN UTILIZATION

The distribution of the energy derived from solar sea power by means of hydrogen rather than by electrical transmission is the subject of this discussion. Hydrogen turns out to be cheaper to transmit over long distances, is easy to store and is a clean burning fuel.

Figure 6.1 shows a schematic of the hydrogen economy. It is made up of three parts: production; transmission, storage and distribution; and utilization. The production of hydrogen can be achieved by various means; nuclear, geothermal or solar sea power production. There are a host of ways of producing hydrogen.

After it is produced it must be transported to the user. Under certain conditions the transmission cost of electricity versus the cost of production and transmission of hydrogen turns out to be favorable to the hydrogen concept. Undoubtedly, if the electric power plant can be located close to its load, it makes little sense to go the hydrogen route. But if one can go directly from thermal energy into hydrogen and oxygen, avoiding the electric conversion step, the picture is different.

Secondly, the production of hydrogen by thermal means (i.e., not involving the electrolysis of water) is being examined. At the European Atomic Energy Commission, they are studying various chemical processes. They have multistep processes where they separate the hydrogen and oxygen and form intermediate compounds and then extract hydrogen and oxygen from these compounds.

FIGURE 6.1: THE HYDROGEN ECONOMY

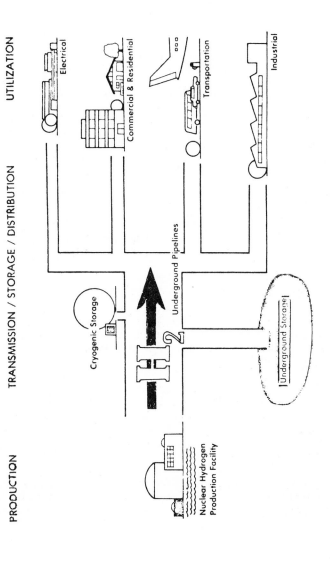

PRODUCTION   TRANSMISSION / STORAGE / DISTRIBUTION   UTILIZATION

Source: PB 228 066

Another reason to go to hydrogen as an intermediate product even when you are next to the generating plant involves the ability to introduce storage. Over-all plant capacity factor of all electrical generating capacity in this country to-day is slightly over 50% because the customer demand curve must be matched. Generating a storable and then using that storable (particularly hydrogen) to cut down the transmission cost and improve the plant capacity factor, is economi-cally more sound than consuming electricity as electricity.

HYDROGEN ENERGY

Solar energy conversion systems conceived of in large-scale typically are viewed as producing electricity as the energy form for general utilization. This discus-sion will focus on hydrogen energy as the product of the ocean-based solar plant. Hydrogen energy embraces hydrogen fuel and/or the hydrogen–oxygen bireactant pair. The latter is produced at the correct chemical ratio (stoichiometric) when-ever the source of the hydrogen energy is solely water.

There is, however, no general agreement as to what to do with the oxygen co-produced from the hydrogen. Opinions vary from releasing it to the atmosphere to carrying it along with the hydrogen for subsequent use in energy conversion.

Hydrogen energy offers significant advantages over electricity. For one thing, hydrogen as a chemical fuel is in concept capable of being substituted for pres-ent day fossil fuels of all kinds. After all, the major share of energy consumed is as chemical fuel, not as electricity. Some sectors of the energy utilization market, namely transportation and electricity generation, are fundamentally de-pendent on chemical energy, a role which hydrogen plays.

Hydrogen offers transportability and storability advantages vis-a-vis electricity. It is cheaper to transmit over very long distances, say 500 miles and more. In concept, hydrogen can be stored as a gas like natural gas, and as a cryogenic liquid like liquefied natural gas (LNG).

Hydrogen is also being examined as a potential fuel for aircraft and automotive transportation systems and for electricity generation, as well as a general syn-thetic fuel for industrial, residential and commercial applications. So, were hy-drogen energy to be the output of future solar energy conversion facilities, it appears that there will be a definite market for this energy form.

Six reasons for considering ocean siting for a solar energy conversion facility are: virtually unlimited area, enormous thermal sink, immediate source of feedstock water, excellent logistics, low-friction bearing surface, and the availability of ocean thermal gradient mode.

Conventionally, large-scale solar energy conversion systems are conceived as land-based, usually in desert locations for reasons of clear skies and low cost, avail-able land area for the extensive solar collector arrays required. The first point, i.e., abundant and available surface area, represents a considerable gain in freedom of geometric location, particularly so for nations located at high latitudes and having characteristically cloudy skies. Also, once the technology of sea-basing has been developed and refined, the environmental sameness of the ocean will allow straightforward standardization in the mechanization of facility hardware.

This will result in important economies, and maximal efficiencies, hence lower energy costs to the eventual consumer of solar energy. Because of the great variability in the terrain and geological makeup of the land, these avenues for long-term cost reductions are not so apparent.

Point two makes note of the fact that all processes attendant to the basic solar energy conversion objective reject heat, and therefore means of cooling must be provided. In other words, energy conversion can only take place if a thermal sink is available, preferably one at low temperatures in order to achieve maximum conversion efficiencies. Usually a source of water must be introduced for these concepts, often in connection with a desalination cofunction.

Thirdly, since hydrogen energy is to be produced in lieu of electricity as an energy carrier, abundant quantities of water for conversion to hydrogen and oxygen will be needed. Although it appears that the ocean's saltwater must be purified prior to being an electrolyzer feedstock, the potential supply of feedstock as such is virtually unlimited. A desalination system in the ocean-based energy complex will be needed perhaps as a solar distillation scheme. This fact leads to consideration of the resulting seawater concentrates as a source of possible coproduction of raw materials (e.g., bromine) and finished goods (e.g., magnesium metal).

Fourth, effective shipping means and deep ocean ports are possible to support the logistical requirements of truly large-scale industrial enterprises on the seas. Waterborne shipping is the cheapest way to move large tonnage products and raw materials. This mode of transport rivals and in many respects exceeds the shipping economies of pipelines for gas and oil.

Point five relates to the flexibility for rotation and translation of an ocean-based energy array. If direct thermal solar collectors of the concentrator type are to be employed, some means of sun-tracking will be necessary. The surface of the ocean can be viewed as a bearing surface with appropriate flotation and propulsion means provided in using the approach of limiting steering motions to only azimuthal rotation.

By virtue of being translatable on the ocean, the solar facility can conceivably be moved in a scheme to maximize solar exposure. For instance, the annual locus of maximal insolation can be programmed and followed by the floating facility as one design option.

Finally, the intrinsic availability of the ocean thermal gradient conversion mode which is the one mode uniquely tied to ocean locations, although shore-based facilities are conceivable also. But indications so far are that practical thermal difference plants will be located in the oceans and not on land.

The ocean-based solar/sea conversion concept is shown in Figure 6.2. The principal constituent systems believed to be required in configuring the overall facility are shown in block-diagram format in Figure 6.2. The bold arrows depict the basic energy production portion of the assemblage. In the dashed box on the left are the solar energy conversion (to electricity) mode alternatives. In one sense these are in competition with one another; on the other hand, there may be an optimal blend of several modes, all powering the electrolyzer system in some time-phased manner.

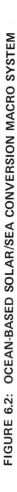
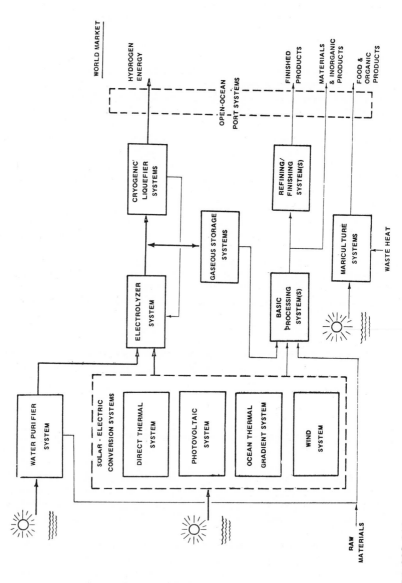

FIGURE 6.2: OCEAN-BASED SOLAR/SEA CONVERSION MACRO SYSTEM

Source: PB 228 066

Previous solar/sea power discussions focussed on the direct thermal-electric approach, using concentrators to heat a working fluid (water) operating in a conventional Rankine turbine-generator system. Acyclic generators were utilized to produce DC power for electrolysis. The objective of this system is in the multimodal conversion manner revealed in Figure 6.2. Perhaps wind power will mutually complement direct thermal; it may turn out that wind availability occurs at times other than when direct sunlight is available for certain ocean locations.

Tracing the macro system makeup further, it is seen that purified seawater is fed to the electrolyzer by a water purifier system. The product hydrogen and oxygen gases, probably generated at high pressure, are stored and subsequently fed to the cryogenic liquefier systems. This is necessary to convert the energy form to a practical transportable form.

Experience with liquefied natural gas would seem to offer a fitting analogy here. Liquefied natural gas is shipped routinely in large capacity oceangoing ships (but not large by oil-carrying supertanker standards). Global capability hydrogen-oxygen cryotankers are envisioned as quite practical delivery means; hence the open ocean port systems, which also will serve the coproduct shipping needs.

Liquefaction of hydrogen energy requires a considerable investment in energy. Based on the higher heating value of hydrogen (an unprecedented 61,000 Btu per pound) as a reference, the liquefaction "energy tax" is of the order of an additional 35 to 40% to liquefy both the hydrogen and oxygen, or about 25% just to liquefy the hydrogen (oxygen used in situ or released to the atmosphere). Alternatives to liquefaction should be sought on one hand, and a means of using the cold at the utilization end should be likewise vigorously pursued in future studies.

Figure 6.3 shows the need for developments in ocean stable platform technology being basic to success in ocean-basing a solar energy conversion macro system of the type and magnitude under consideration. The left hand side attempts to estimate (1) the location above and below the ocean surface, and (2) the relative massiveness of associated hardware for each of the constituent systems shown in Figure 6.2.

Shown is a vertical arrangement with wind power generation equipment at the top, perhaps 100 meters above the ocean surface, and the depth-located electrolyzer and gas storage containment at the bottom. The depth required is a function of system operating pressure levels and the degree of hydrostatic pressure-balancing employed. The indicated 1,000 meters would be compatible with hydrogen and oxygen pressures of about 1,500 pounds per square inch, a nominal number.

Note that the majority of the systems are located on/near the ocean surface. On the right side of Figure 6.3, an initial conceptualization of how the hardware might be arranged is offered. Two basic stable platform types are shown: (1) a relatively compact semisubmersible unit for the industrial system types, e.g., liquefiers, basic processing plant, and (2) a spar buoy configuration, to be repeated periodically for supporting the elevated solar collectors, and the depth-located systems.

FIGURE 6.3: TYPICAL SYSTEMS LOCATION

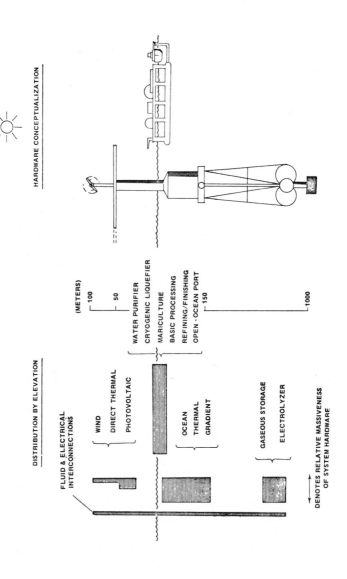

Considerable experience is available to draw upon for both configurations, the first from the offshore oil drilling rig developments, and the second from proposed oceanographic research vessels such as MOSES (Manned Open Sea Experimentation Station), a project at the Oceanic Institute.

NITINOL UTILIZATION

The U.S. Naval Ordnance Laboratory developed a series of engineering alloys that possess a unique mechanical (shape) memory. The generic name of the series of alloys is 55-Nitinol, where Nitinol stands for Nickel Titanium Naval Ordnance Laboratory. These alloys, which have chemical compositions in the range from about 53 to 57 weight percent nickel, balance titanium, are based on the intermetallic compound NiTi.

The memory is such that, given the proper conditions, Nitinol objects can be restored to their original shape even after being "permanently" deformed out of that shape. The return to the original shape is triggered by heating the alloy to a moderate temperature. Considerable force is exerted, and mechanical work can be done, by the material as it snaps back to its original shape. This mechanical (shape) memory, which is otherwise unknown in engineering alloy systems, furnishes design engineers with the opportunity to design on the basis of an entirely new principle.

The concept of a shape memory is new. Accordingly, a brief description of this memory and how it is imparted to Nitinol objects will be given first. The steps in the Nitinol shape memory process are shown in Figure 6.4. The material is first obtained in a basic shape such as wire, rod, sheet, tube, extrusion, and casting (step 1). This shape is then formed into the shape that the alloy will later be called upon to remember, i.e., its memory configuration (step 2).

Next, the Nitinol shape is clamped in a fixture that constrains it in the memory configuration (step 3). The Nitinol, restrained from moving by the fixture, is given a heat treatment to impart the memory and is then cooled (step 4). 900°F was found to be the optimum heat treatment temperature for several 55-Nitinol compositions.

After the Nitinol part, in the memory configuration, has cooled to below the transformation temperature range (to be defined), the part is strained to form the intermediate shape (step 5). The intermediate shape is the shape that the part is to retain until it is heated to restore it to the memory configuration (step 6). The temperature to which the part must be heated to return it to the memory configuration depends on the chemical composition of the alloy.

The transformation temperature on heating, which corresponds approximately to the top of the temperature range through which the material must be heated to restore it to its memory configuration, varies in binary nickel-titanium alloys from about -55° to +166°C. By substituting cobalt for some of the nickel in the alloy, the transformation temperature can be decreased to about -238°C. Of considerable interest to potential users of the alloy is the fact that the memory process (step 5 through 7) can be repeated many times. That is, after the part has recovered its memory configuration upon heating (step 6) and cooled to below its transformation temperature range (step 7), it can be deformed again to an intermediate shape (step 5) and then heated to restore it to the memory configuration (step 6). This repeatability of the shape-memory effect has been demonstrated on samples that have been subjected to steps 5, 6, and 7, thousands and even millions of times.

As the 55-Nitinol part, in its intermediate shape, is heated to return it to its memory configuration, the alloy exerts very considerable force and can do significant mechanical work. It is this property that allows the design of an SSPP Nitinol engine. Figure 6.5 shows that the recovery at the 6% constant strain level is practically constant and does not fall off significantly after many cycles.

FIGURE 6.4: STEPS IN THE NITINOL SHAPE MEMORY PROCESS

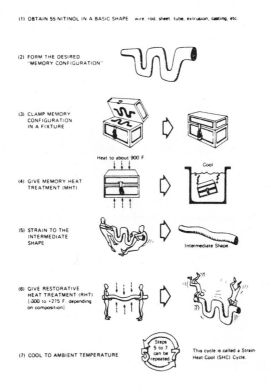

Source: NASA SP 5110

Early tests showed that 55-Nitinol was not corroded during a 96 hour salt spray test (35°C, 20% salt solution). Neither was it attacked by seawater, a normal air atmosphere, or during normal handling. Additional experiments to evaluate the resistance of Nitinol alloys to a seawater environment involved studies of the resistance of Nitinol alloys to (1) impingement by high velocity seawater, (2) cavitation erosion, (3) stress corrosion, and (4) crevice corrosion.

FIGURE 6.5: SHAPE RECOVERY FATIGUE CURVE FOR 0.020-INCH DIAMETER NITINOL WIRE

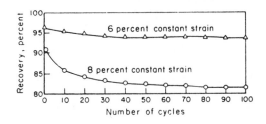

Source: NASA SP 5110

In the high-velocity impingement tests, commercial-purity specimens of 55-Nitinol (55 and 56 weight percent nickel) were subjected to impingement by seawater at 15 to 26 fps for a period of 60 days. The weight loss varied from nil to a few milligrams; the size of the specimens was not given. After testing, all specimens retained the appearance that they had prior to testing.

Rectangular radius-nosed bars with transverse holes downstream were used for the cavitation-erosion tests; once again, commercial-purity alloys containing about 55 and 56 weight percent Ni were used. The specimens were subjected to seawater at 117 fps for a period of 30 days. At the end of this period, the 55 weight percent Ni alloy had lost about 0.092 gram and the 56 weight percent Ni alloy had lost 0.050 gram.

The susceptibility of Nitinol (55 weight percent Ni) to stress corrosion in seawater was measured on bar specimens containing a fatigue crack at the root of a V-notch. Such specimens, stressed at up to 100% of their yield stress while immersed in seawater, showed no tendency toward the propagation of the crack. The duration of the test was not noted.

The ability of Nitinol to do work while being restored to its memory shape is of great practical significance. Essentially, Nitinol acts as a transducer, converting thermal energy or electrical energy to mechanical energy. For example, Nitinol is the basis of a method described by W.J. Buehler and D.M. Goldstein in U.S. Patent 3,403,238 for converting heat energy to mechanical energy. Figure 6.6 illustrates the principle.

A strip of Nitinol is deflected, at the subsurface (cold) water temperature in the SSPP, by an amount Z by a weight W_1. A second weight W_2 is added after the maximum deflection Z is attained. The SSPP surface water (warm) then is brought in to flow over the Nitinol. The temperature of the surface water is sufficient to exceed the Nitinol transition range temperature and return the

Nitinol to its zero-deflection or historical configuration. The amount of useful work performed ($W_1Z + W_2Z$) is greater than the work used to deform the Nitinol initially (W_1Z). Thus the Nitinol strip has converted the thermal energy of the ocean into mechanical energy which can then be converted into electrical energy by standard procedures.

FIGURE 6.6: CONVERSION OF HEAT ENERGY TO MECHANICAL ENERGY USING 55-NITINOL

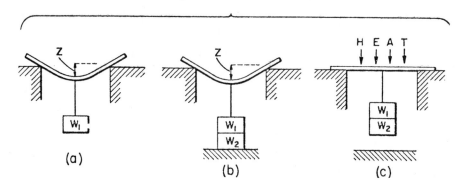

Source: U.S. Patent 3,403,238

The fatigue strength of Nitinol is high both in the pre- and post-transition temperature ranges and as shown in Figure 6.5, there is no significant decay in percent recovery with increasing cycles. The data also indicates that one should not use strains above 8% insofar as this strain appears to be in the vicinity of the maximum recovery stress and work. Efficiencies of 25% have been obtained and it is hoped that SSPP Nitinol can be designed to operate appropriately and highly efficiently at temperature differences one observes at the prime SSPP sites. From preliminary calculations it appears that Nitinol engines will cost less than the equivalent closed cycle systems proposed.

FURTHER SSPP PROPOSALS

Ocean Thermal Energy Conversion (OTEC)

Ocean thermal energy conversion using temperature extremes of 40°F is adequate for producing electricity, as determined by Lockheed engineers who conducted a study for the National Science Foundation. The floating plants can be built from existing components, modified to meet OTEC requirements.

In the study, a baseline model of a generating plant that could send 160 million watts of power ashore to distribution networks was formulated. Minor improvements to the baseline could produce busbar electricity at a cost of 27 mills per kilowatt hour. Technical advances could reduce the cost.

The closed-cycle concept would heat a liquid until it becomes a gas under pressure and drives a turbine hooked to a generator. Ammonia is proposed as the working fluid. After passing through the turbine, the gas is cooled by water from the ocean depths until it becomes liquid and begins the cycle again.

Under the present concept the generating unit would be 1,500 feet long with a diameter of 246 feet. The 260,000 ton concrete platform would be nearly 600 feet from top to bottom with 35,000 tons of concrete telescoping cold water pipe extending about 1,000 feet into the ocean depths.

A knob on top of the platform extending 60 feet above the ocean surface would serve as a helicopter pad and support ventilator systems. The platform would provide crew quarters, an ammonia tank, and electrical connections to shore. Removable steel and concrete modules housing the turbine-generators, pumps and heat exchangers would be attached to the outside of the platform.

Danish System

It has been reported that a proposal for extracting energy from the sea was made by Earl J. Beck of the energy program office at the Naval Construction Battalion Center at Port Hueneme, California. It has been elaborated by scientists of Carnegie-Mellon University in Pittsburgh.

It differs basically from other such plans, although it depends on the temperature difference between frigid bottom water and the warm surface water of low latitudes.

The plan is related to the air lift pump sometimes used where water is too silty to pass through a mechanical system. Compressed air is squirted into the bottom of a pipe deep in the water. It expands, forming a bubbly air-water mixture that, in a confined system, is light enough to rise some distance above the water's surface.

In Mr. Beck's plan cold water is lifted from the sea floor, probably several thousand feet. Near the surface water warmer by some 40°F is introduced into the cold water via a confined aperture. This increases its velocity and reduces its press pressure sufficiently so that the warm water begins to boil.

Theoretically, the resulting mixture of bubbles and water in what Mr. Beck calls his steam lift water pump could rise hundreds of feet, he said. The water would then run downhill to drive a turbine. If the power plant was afloat, he wrote, a tower several hundred feet high might be impractical, but if the system was closed, a lesser lifting of the water would be sufficient, since the pressure of the steam bubbles would contribute to the driving power.

The Carnegie-Mellon modification would introduce foam, which would be more efficient in enabling the steam bubbles to carry water upward. In this way, according to the calculation, the water could be lifted 648 feet above sea level in 21 seconds.

The Carnegie-Mellon contribution was by Dr. Clarence Zener, who for several years has been working on plans for extracting ocean energy, and Dr. John Fetkovich.

Japanese System

The Japanese are also interested in the design of an ocean-based SSPP. Shown in Figure 6.7 is a preliminary conceptual design for a 1500 kw$_e$ SSPP. The specifications for the SSPP are shown in Table 6.1 for two working fluids. The cost data for this SSPP with respect to these working fluids is shown in Table 6.2.

TABLE 6.1: SPECIFICATION OF THE EXPERIMENTAL 1500 kw SSPP

Working Fluids / Items	NH$_2$	R-114
Turbine and Generator		
Generator output (kw)	1,500	1,500
Turbine (RPM)	3,600	3,600
Blade height (mm)	~60	~410
Working fluid flow (ton/h)	173	1,270
Working fluid volume (ton)	220	490
Evaporator and Condenser		
Evap.-heat exchange (G cal/h)	50.8	43.5
Cond. heat exchange (G cal/h)	49.5	42.0
Warm water temp (°C)	29	29
Cold water temp (°C)	7	7
Intake water		
Warm side (m³/h)	25,900	22,200
Cold side (m³/h)	25,200	21,400

Working Fluids / Items	NH	R-114
Intake Equipments		
Pipes (FRP)	1 m ⌀ x 9 (2 m ⌀ x 2)	1 m ⌀ x 9 (2 m ⌀ x 2)
Cold pipe length (m)	1,000	1.000
Pump power (kw)		
Cold side	330	300
Warm side	120	110
Working fluid	40	30
Civil Works		
SSPP plant (m²)	150	150
Concrete works (m³)	950	950
Sub-merged		
Concrete works (m³)	500	500
FRP anchor (ton)	50	50
FRP works (man. day)	300	300
Others		

TABLE 6.2: COST ESTIMATION OF EXPERIMENTAL 1500 kw SSPP

Cost / Working Fluids	NH$_3$	R-114
Turbine and Generator	2.2 x 10⁵$	2.3 x 10⁵$
Evaporator and Condenser (shell and tube type)	9.7 x 10⁵$	10.8 x 10⁵$
Pumps and Motors	2.4 x 10⁵$	2.6 x 10⁵$
Intake pipes (FRP)	15.3 x 10⁵$	15.3 x 10⁵$
Electric Equipments	1.7 x 10⁵$	1.7 x 10⁵$
Water Pipe Screen	0.3 x 10⁵$	0.3 x 10⁵$
Working Fluid/Pump	0.2 x 10⁵$	0.2 x 10⁵$
Working Fluid	0.4 x 10⁵$	12.7 x 10⁵$
Civils	10.4 x 10⁵$	0.4 x 10⁵$
Reserve	2.1 x 10⁵$	3.3 x 10⁵$
Total Construction Cost	44.6 x 10⁵$	59.1 x 10⁵$
Generated Power per Year	7.53 x 10⁶kwh	7.9 x 10⁶kwh
Unit Construction Cost	3.000 $/kw	3.900 $/kw
Unit Power Cost (at transmission terminal)	89 mill/kwh	112 mill/kwh

Source: NASA TM X 70783

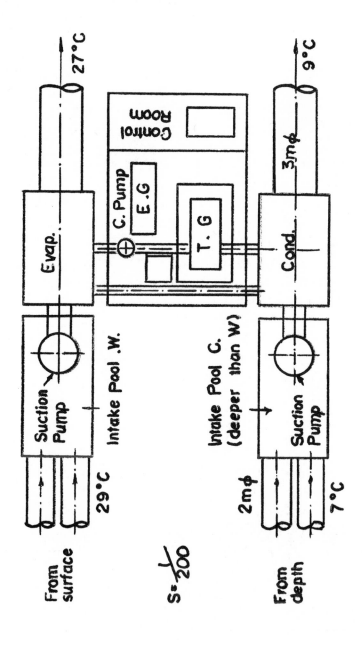

FIGURE 6.7: PLAN OF A 1500 kw EXPERIMENTAL SSPP

This plant is designed to be placed in tropical waters. It is assumed that the surface waters are at a temperature of 29°C. The cold water at the intake pipe is assumed to be at 7°C. Note that small changes in the cold water temperature produce relatively large changes in power production as shown in Table 6.3.

TABLE 6.3: DEPENDENCY OF POWER OUTPUT ON COLD WATER TEMPERATURE

Cold water temperature		7°C		8°C		9°C	
Items Working Fluids		NH_3	R-114	NH_3	R-114	NH_3	R-114
Output power (kw) (at the generator terminal)		1.500	1.500	1.380	1.410	1.250	1.320
Unit power cost (mill/kwh) (at the transmission terminal)		89	112	97	114	113	124

Source: NASA TM X 70783

REFERENCES

PB 228 066   *Proceedings, Solar Sea Power Plant Conference and Workshop,* A. Lavi, Carnegie-Mellon University, June 27-28, 1973.

PB 228 067   *Research Applied to Ocean Sited Power Plants,* J.H. Anderson, Massachusetts University, January 25, 1974.

PB 228 068   *Solar Sea Power,* C. Zener, et al, Carnegie-Mellon University, January 25, 1974.

PB 228 069   *Solar Sea Power,* C. Zener, et al, Carnegie-Mellon University, October 11, 1973.

PB 228 070   *Research Applied to Ocean Sited Power Plants,* J.H. Anderson, Massachusetts University, July 31, 1973.

PB 231 142   *Systems Analysis of Solar Energy Programs,* Mitre Corporation, December, 1973.

PB 235 469   *Solar Sea Power,* C. Zener, et al, Carnegie-Mellon University, April 30, 1974.

PB 236 422   *Technical and Economic Feasibility of the Ocean Thermal Differences Process as a Solar-Driven Energy Process,* The Energy Alternatives Program, University of Massachusetts, July 31, 1974.

PB 238 571   *Feasibility Study of a 100 Megawatt Open Cycle Ocean Thermal Difference Power Plant,* J.L. Boot and J.G. McGowan, University of Massachusetts, August, 1974.

PB 238 572   *Variations in Heat Exchanger Design for Ocean Thermal Difference Power Plants,* J.G. McGowan, et al, Massachusetts University, August, 1974.

PB 239 369   *Heat Exchanger for Sea Solar Power Plants,* J. Hilbert Anderson, Massachusetts University, September, 1973.

PB 239 371   *Ocean Thermal Difference Power Plant Turbine Design,* L.L. Ambs and J. Marshall, University of Massachusetts, Nov. 1973.

PB 239 373   *Technical and Economic Evaluation of Ocean Thermal Difference Power Plant Turbomachinery,* R.D. Lessard, University of Massachusetts, December, 1973.

PB 239 393   *Preliminary Investigation of an Open Cycle Ocean Thermal Difference Power Plant Design,* J. Boot, et al, Massachusetts University, August, 1973.

NASA TM X 70783   *Solar Sea Power Plants (SSPP),* September, 1974.

NASA SP 5110   *55-Nitinol—The Alloy With a Memory: Its Physical Metallurgy, Properties, and Applications,* 1972.

AD 779877   *Energy from the Ocean: An Appraisal,* O.M. Griffin, Naval Research Laboratory, May, 1974.

OCEAN WASTE DISPOSAL PRACTICES 1975

by Alexander W. Reed

Pollution Technology Review No. 23
Ocean Technology Review No. 4

This book is an evaluation of waste discharges and practices as they affect the littoral regions of the United States, with many projections of the ultimate influence upon the world ocean.

The problems of rapidly increasing waste loads arising from the exponential increase in populations make a constant review of the assimilation capacity of coastal waters a pressing necessity. The overall problem is not a static one. The level of research and study must keep pace in order to meet the growing demand for transforming the wastes into an assimilable condition before dumping them into the ocean.

Waste generation is inevitable—most wastes eventually reach the ocean whether they are buried on land, discharged into the air or run into streams. But there is now a growing public awareness that the ocean must be kept reasonably intact not only as a means for marine transport, but also to develop heretofore untapped resources such as unfamiliar but abundant fish species, offshore sand and gravel, deep water manganese deposits, thermal gradients and currents and artificially induced upwellings to produce electrical energy. Clean seawater, beaches, and shorelines are valuable resources and indispensable to mankind for their recreational and aesthetic values.

The major purposes of the government-sponsored reports from which this book is constructed is to provide scientific findings, evaluations, and conclusions with respect to possible long-range effects of oceanic pollution and man-induced changes of ocean ecosystems. This book should help to weigh these effects against the need for and benefits to be derived from ocean dumping and its alternatives. A partial and condensed table of contents follows here.

1. THE CASE FOR OCEAN DUMPING

2. DEFINING OCEAN WASTES & ASSESSING THE EFFECTS
Parameters of Coastal Waste Disposal
Physical Description
Location of Disposal Areas
Disposal Methods, Tonnage, Costs
Environmental Effects
Beneficial Uses of Solid Wastes
Dredge Spoil
Global Marine Ecosystems

3. SCIENTIFIC PARAMETERS NEEDED FOR PLANNING DISPOSAL IN OCEAN WATERS
Studying the Marine Environment
Criteria for Disposal of Dredged Material
Criteria for Selecting a Disposal Site
Coastal Waste Management
Transport Mechanisms
Motion Studies
Dispersion Studies
Biological Processes
Effects on Receiving Waters
Eutrophication: Nitrogen as Factor
Chemical & Geochemical Processes

4. MONITORING THE DISPOSALS
Operational Aspects
Monitoring Equipment
Navigational Aids
Information & Surveillance Systems
Using Satellites (ERTS)
Control Criteria

5. ECONOMIC ASPECTS
Using Barges
Dumping of Refuse
Incineration at Sea
Municipal Waste Disposal
   by Shipborne Incineration
Ship Incinerator Costs
Off-Shore Dumpsites for
   Incinerator Residues

6. NEW YORK & MIDATLANTIC BIGHT STUDIES
Case Histories
Estuarine Economics
EPA Facts and Figures
Environmental Findings 1974/75
Assessing Ocean Pollutants
Environmental Survey of a
   Midatlantic Bight Dumpsite

7. OTHER GEOGRAPHIC AREAS
Charleston Study
Gulf Coast Study
Southern California Study
Puget Sound Study
Analysis of Dumping Activities
Recommended Procedures

ISBN 0-8155-0591-4

334 pages